USING THE BUILDING REGULATIONS

Site Preparation and Resistance to Contaminants and Moisture

MJ Billington

ELSEVIER

AMSTERDAM • BOSTON • HEIDELBERG • LONDON • OXFORD
NEW YORK • PARIS • SAN DIEGO • SAN FRANCISCO
SINGAPORE • SYDNEY • TOKYO

Butterworth-Heinemann is an imprint of Elsevier
Linacre House, Jordan Hill, Oxford, OX2 8DP
30 Corporate Drive, Burlington, MA 01803

First edition 2007

British Library Cataloguing in Publication Data
A catalogue record for this book is available from the British Library

Library of Congress Cataloging-in-Publication Data
A catalog record for this book is available from the Library of Congress

ISBN 13: 9-78-0-75-066258-1

For information on all Butterworth-Heinemann publications visit our
web site at www.books.elsevier.com

Typeset by CEPHA Imaging Pvt Ltd, Bangalore.
Printed and bound in Great Britain

07 08 09 10 11   10 9 8 7 6 5 4 3 2 1

Author biography

M J Billington is a Chartered Building Surveyor. He has a lifetime's involvement in the construction industry having worked at one time or another in design, construction and control (both private and public sectors). He was formerly Senior Lecturer in building control and construction at De Montfort University, Leicester, before leaving to join the private sector, where he continued to act as visiting lecturer at a number of universities. He has published many technical papers and a number of books on building regulations and building defects, and is a contributor to Knights Guide to Building Control Law and Practice. Currently, he is Managing Director of Construction Auditing Services Limited, a company that specializes in latent defects insurance technical auditing, and is Managing Director of Centass Ltd, a Building Regulations Competent Person Scheme for replacement windows and doors.

Contents

List of figures

List of tables

Preface

Unfortunately, it is a fact that the Building Regulations and Approved Documents get more complex with every update, often requiring the services of specialist professionals to make sense of the provisions (which even they find difficult to understand!). New areas of control are being introduced each year and the scope of the existing regulations is being extended with each revision.

The current Approved Documents only provide detailed guidance in the design of extremely simple and straightforward buildings using mainly traditional techniques. Larger and more complex buildings require access to other source documents but, although these are referred to in the Approved Documents, no details of their contents or advantages of use are given. Therefore, the Approved Documents are becoming less and less useful to anyone concerned with the design and construction of any but the most simple of buildings, and are in danger of becoming merely an index of references.

To date, currently published guidance works on the Building Regulations and Approved Documents have tended merely to restate the official guidance in a simpler fashion with additional illustrations. Some of these works have restructured the guidance in an attempt to make it more user-friendly, but have not really added value. Additionally, where the Approved Documents have made reference to alternative guidance sources the texts have tended to do the same – without attempting to state what advantages might be gained by using a different approach and without giving even the most basic details of what might be contained in the reference texts.

In early 2003, the author approached the publishers of this book, Elsevier, with an idea for a new kind of building control guidance publication. The aim would be to present a series of books, each one covering a separate Approved Document, which would provide users with much more detailed guidance than the Approved Documents currently provide. Furthermore, these would be written by experienced engineers, surveyors, architects and building control surveyors, etc., who had relevant specialist knowledge and hands-on experience.

This book is the third in the series and concentrates on Part C and its accompanying Approved Document C dealing with site preparation and resistance to contaminants and moisture. The text covers the latest amendment to Part C which came into force on 1 December 2004.

Part 1 of this book is introductory and gives a broad overview of the Building Regulations and the control system. It also includes an introduction to Part C and Approved Document C and sets out the factors which were considered in the revision of the previous (1992) edition of Approved Document C.

Part 2 covers site preparation and resistance to contaminants. The guidance on resistance to contaminants has undergone a great deal of revision, due mostly to new regulations made under Part IIA of the Environmental Protection Act 1990. This has resulted in the provision of new guidance on the risk assessment of sites.

Part 3 deals with resistance to moisture. Here, the most significant changes involve new guidance on avoiding the occurrence of interstitial condensation in floors, walls and roofs, and on measures to prevent surface condensation and mould growth. To achieve this, some of the guidance that was previously located in Approved Document F Ventilation (concerning ventilation of flat and pitched roofs) has been updated and relocated in Approved Document C.

The aim of this book (and this series of books) is to provide a convenient, straightforward, comprehensive guide and reference to a complex and constantly changing subject. It must be stressed that the books in this series are a guide to the various regulations and approved and other documents, but are not a substitute for them. Furthermore, the guidance in the Approved and other source documents is not mandatory and differences of opinion can quite legitimately exist between controllers and developers or designers as to whether a particular detail in a building design does actually satisfy the mandatory functional requirements of the Building Regulations.

The intended readers are all those concerned with building work – architects and other designers, building control officers, approved inspectors, building surveyors, clerks of works, services engineers, contractors, and teachers in further and higher education, etc. – as well as their potential successors, the current generation of students.

The law in this book is stated on the basis of cases reported and other material available to us after 1 November 2006.

MJ Billington

Acknowledgements

The author wishes to thank the editorial and production team at Elsevier for their help, patience and dedicated support in the publication of this book.

About this series of books

Whether we like it or not, the Building Regulations and their associated government-approved guidance documents get more complex with every update, often requiring the services of specialist professionals (services engineers, fire engineers, etc.) to make sense of the provisions. New areas of control are being introduced each year and the scope of the existing regulations is being extended with each revision.

The technical guidance given in the current Approved Documents is only of use in the design of extremely simple and straightforward buildings using mainly traditional techniques. For larger and more complex buildings it is usually better (and more efficient in terms of building design) to use other sources of guidance (British and European Standards, Building Research Establishment Reports, etc.) and although a great many of these other source documents are referenced in the Approved Documents no details of their contents or advantages of use are given.

The current Approved Documents usually fail to provide sufficient guidance just when it is needed, that is, when it is proposed to deviate from the simple solutions or attempt to design something slightly unusual, thus encouraging adherence to traditional and (perhaps) unimaginative designs and details, and discouraging innovation in the majority of building designs.

This series of books, by addressing different parts of the Building Regulations in separate volumes, will enable each Part to be explored in detail.

The information contained in the Approved Documents is expanded not only by describing the traditional approach but also by making extensive reference to other sources of guidance contained in them. These 'alternative approaches' (as they are called in the Approved Documents) are analysed and their most critical parts are presented in the text with indications of where they can be used to advantage (over the traditional approach).

As this is a new concept in building control publications our aim is to develop the series by including examples of radical design solutions that go beyond the Approved Document guidance but still comply with the Regulations. Such innovative buildings already exist, one example being the Queen's Building at De Montfort University in Leicester, which makes extensive use of passive stack ventilation instead of traditional opening windows or air conditioning.

About this book

This book presents a detailed analysis of Part C – Site preparation and resistance to contaminants and moisture – of Schedule 1 to the Building Regulations 2000 and its accompanying Approved Document. Approved Document C became operative on 1 December 2004 and essentially, replaced the former 1992 edition which related to the Building Regulations 1991.

In deciding to revise this Part of the Regulations the Government identified three options:

Option 1. Do nothing.
Option 2. Carry out minimal changes involving a simple update of the references in Approved Document C and publicize good practice.
Option 3. Substantially amend the Building Regulations and the technical guidance in Approved Document C.

It is probably no surprise that they chose the third option! An outline of the amendments is given in Chapter 2.

When the new Approved Document was published is was immediately apparent that not only had it increased greatly in size, but the number of references to other sources of information had also greatly increased – from 35 separate references in the 1992 edition to 110 in the 2004 edition.

From the viewpoint of a user (who might or might not be a specialist), the questions that arise from allowing so many references to appear in what is supposed to be a technical guidance document are:

- How relevant are each of the references to me?
- Do I need to buy them all, or just some of them?
- If I need to buy some, which ones should I get?

We have tried to address these questions in this book by reviewing all the references where possible (and some of them are over 300 pages in length) and by presenting these reviews in TWO places:

- within the text where they appear logically, in line with the Approved Document guidance, and
- In a summarized version in Appendix 1, in order to give a quick reference source.

After carrying out these reviews it is apparent that the references can be divided into three categories:

(a) those that are non-technical and are of universal application
(b) those that are best suited to building professionals (designers, surveyors, contractors, etc.) because of their technical content
(c) those that are needed only by specialists (soil and geotechnical engineers, building control bodies, Environment Agency staff, etc.) because of the specialist nature of their contents.

Our aim is that these analyses should help readers make an informed choice as to which of the reference documents they really need and in this way save time and money.

This book is aimed at designers, builders, students on construction- and building-related surveying courses, building control professionals and anyone else with an interest in the built environment. Its purpose is to keep them better informed and more able to deal with a complex and evolving area of law which directly affects everyone.

PART 1

Introduction

1

Series introduction

1.1 Introduction

Although we may not be aware of it, the influence of the Building Regulations is around us all of the time.

In our homes building regulations affect and control the:

- size and method of construction of foundations, walls (both internal and external), floors, roofs and chimneys
- size and position of stairs, room exits, corridors and external doors
- number, position, size and form of construction of windows and external doors (including glazing)
- methods for disposing of solid waste
- design, construction and use of the services such as:
  - above and below ground foul drainage taking the waste from kitchen and bathroom appliances (including the design and siting of the appliances themselves)
  - rainwater disposal systems including gutters and downpipes from roofs and drainage from paths and paving
  - electrical installations
  - heating and hot water installations using gas, oil or solid fuel
  - fire detection and alarm systems
  - mechanical ventilation systems
- design and construction of the paths outside the house that:
  - lead to the main entrance, and
  - are used to access the place where refuse is stored.

In a similar manner, they also affect the places where people go when away from their homes such as:

- factories, offices, warehouses, shops and multi-storey car parks
- schools, universities and colleges
- leisure, sport and recreation centres
- hospitals, clinics, doctors surgeries, health care centres and other health care premises
- hotels, motels, guest houses, boarding houses, hostels and halls of residence
- theatres, cinemas, concert halls and other entertainment buildings
- churches and other places or worship.

In fact, anything that can normally be considered to be a building will be affected by building regulations. But it is not just the design and construction of the building itself that is controlled.

The regulations also affect the site on which the building is placed in order to:

- lessen the effect of fire spread between neighbouring buildings
- permit access across the site for the fire brigade in the event of fire
- allow access for disabled people who may need to get from a parking place or site entrance to the building
- permit access for refuse collection.

1.2 What are the Building Regulations?

When asked this question most people (assuming that they have even heard of the regulations) will usually bring to mind a series of A4 documents with green and white covers and the words 'Approved Document' on the front! These documents are not, of course, the Building Regulations, but have come to be regarded as such by most builders, designers and their clients, and it is this misconception that has led to a great deal of confusion regarding the true nature of the building control system and the regulations. When applied to England and Wales, the Building Regulations consist of a set of rules that can only be made by Parliament for a number of specific purposes. The purposes include:

- ensuring the health, safety, welfare and convenience of persons in or about buildings and of others who may be affected by buildings or matters connected with buildings
- furthering the conservation of fuel and power
- preventing waste, undue consumption, misuse or contamination of water.

The regulations may be made 'with respect to the design and construction of buildings and the provision of services, fittings and equipment in or in connection with buildings'.

Originally (in Victorian times), the regulations (or byelaws as they were known then) were concerned only with public health and safety, but in the late twentieth century additional reasons for making building regulations were added so that it would now seem possible to include almost anything under the banner of 'welfare and convenience'.

The Regulations are of two types:

(a) Those that deal with issues of procedure or administration such as:
   - the types of work to which the regulations apply
   - the method of making an application to ensure compliance and the information that must be supplied to the controlling authority
   - the frequencies and stages at which the control authorities must be informed of the work
   - details of the testing and sampling that may be carried out by the controlling authorities to confirm compliance
   - what sorts of work might be exempted from regulation control
   - what can be done in the event of the work not complying with the regulations.
(b) Those that describe the 'standards' which must be met by the building (called 'substantive' requirements) such as:
   - the ability of the building to:
     - retain its structural integrity
     - resist the effects of fire and allow people to escape if a fire should occur
     - resist dampness and the effects of condensation
     - resist the passage of sound
     - minimize the production of carbon dioxide by being energy efficient
     - be safe to use, especially where hazards of design or construction might exist, such as on stairways and landings or in the use of glass in windows, doors or as guarding
     - maintain a healthy internal environment by means of adequate ventilation.
   - the safe installation and use of the building's services including:
     - electric power and lighting
     - boilers, open fires, chimneys, hearths and flues
     - unvented heating and hot water systems
     - sanitary installations and above and below ground drainage
     - foul and waste disposal systems
     - mechanical ventilation and air conditioning systems
     - lifts and conveyors.

Because the regulations are phrased in functional terms (i.e. they state what must be achieved without saying how this must be done) they contain no practical guidance regarding methods of compliance. The intention of this approach is that it gives designers and builders flexibility in the way they comply and it does not prevent the development and use of innovative solutions and new materials and methods of construction. Of course, much building work is done in traditional materials using standard solutions developed over many years and based on sound building practice. To assist designers and contractors in these accepted methods the government has provided non-mandatory guidance principally in the form of 'Approved Documents', there being an Approved Document that deals with each substantive provision of the Building Regulations. This does not prevent the use of other 'official' documents such as and Harmonised Standards (British or European), and the adoption of other methods of demonstrating compliance such as, past experience of successful use, test evidence, calculations, compliance with European Technical Approvals, the use of CE-marked materials, etc.

1.3 How are the Regulations administered?

For most types of building work (new build, extensions, alterations and some use changes) builders and developers are required by law to ensure that they comply with the Regulations. At present this must be demonstrated by means of an independent check that compliance has been sought and achieved.

For this purpose, building control is provided by two competing bodies – local authorities and Approved Inspectors.

Both building control bodies will charge for their services. They may offer advice before work is started, and both will check plans of the proposed work and carry out site inspections during the construction process to ensure compliance with the statutory requirements of the Building Regulations.

1.3.1 Local Authority Building Control (LABC)

Each local authority in England and Wales (unitary, district and London boroughs in England and county and county borough councils in Wales) has a Building Control section. The local authority has a general duty to see that building work complies with the Building Regulations unless it is formally under the control of an Approved Inspector.

Individual local authorities co-ordinate their services regionally and nationally (and provide a range of national approval schemes) via LABC Services.

Full details of each local authority (contact details, geographical area covered, etc.) can be found at www.labc-services.co.uk.

1.3.2 Approved Inspectors

Approved Inspectors are companies or individuals authorized under Sections 47 to 58 of the Building Act 1984 to carry out building control work in England and Wales.

The Construction Industry Council (CIC) is responsible for deciding all applications for Approved Inspector status. A list of Approved Inspectors can be viewed at the Association of Consultant Approved Inspectors (ACAI) web site at www.acai.org.uk.

Full details of the administrative provisions for both local authorities and Approved Inspectors may be found in Chapter 5 of the accompanying book in this series *Using the Building Regulations – Administrative Procedures*, first edition 2005 (published by Elsevier Butterworth Heinemann ISBN 0 7506 6257 3).

1.4 Why are the Building Regulations needed?

1.4.1 Control of public health and safety

The current system of building control by means of government regulation has its roots in the mid-Victorian era. It was originally set up to counteract the truly horrific living and working conditions of the poor working classes who had flocked to the new industrial towns in the forlorn hope of making a better living. Chapter 2 of the first book in this series (*Using the Building Regulations – Administrative Procedures*, details as above) describes the factors which caused this exodus from the countryside and the conditions experienced by the incomers; factors which led to overcrowding, desperately insanitary living conditions and the rapid outbreak and spread of disease and infection. There is no doubt that a punitive system of control was needed at that time for the control of new housing, and the enforcement powers given to local authorities (coupled with legislation that dealt with existing sub-standard housing) enabled the worst conditions to be eradicated and the spread of disease to be substantially halted.

The Victorian system of control based purely on issues of public health and safety enforced by local authorities continued to be effective for the next 100 years, the only major change being the conversion of the system from local byelaws to national regulations in 1966.

1.4.2 Welfare and convenience and other controls

The first hint of an extension of the system from one based solely on public health and safety came with the passing of the Health and Safety at Work etc. Act in 1974 (the 1974 Act). Part III of the 1974 Act was devoted entirely to changes in the building control system and regulations, and it increased the range of powers given to the Secretary of State. Section 61 of the 1974 Act enabled him to make regulations for the purposes of securing the welfare and convenience (in addition to health and safety) of persons in or about buildings. Regulations could also be made now for furthering the conservation of fuel and power and for preventing the waste, undue consumption, misuse or contamination of water. The 1974 Act was later repealed and its main parts were subsumed into the Building Act 1984.

1.4.3 The new system and the extension of control

Initially, the new powers remained largely unused and it was not until the coming into operation of the completely revamped building control system brought about by the 1984 Act and the Building Regulations 1985 that the old health and safety based approach began to change. The 1984 Act also permitted the building control system to be administered by private individuals and corporate (i.e. non-local authority) bodies called Approved Inspectors in competition with local authorities, although enforcement powers remained with local authorities. The new powers have resulted in the following major extensions of control to:

- heating, hot and cold water, mechanical ventilation and air conditioning systems
- airtightness of buildings
- prevention of leakage of oil storage systems
- protection of LPG storage systems
- drainage of paths and paving
- access and facilities for disabled people in buildings (although the reference to disabled people has now been dropped)
- provision of information on the operation and maintenance of services controlled under the regulations
- measures to alleviate the effects of flooding in buildings
- measures to reduce the transmission of sound within dwellings and between rooms used for residential purposes in buildings other than dwellings.

Furthermore, a consultation in 2004 put forward proposals intended to facilitate the distribution of electronic communication services (broadband) around buildings in a proposed Part Q, presumably under the banner of convenience.

As the scope of control has increased, the government has attempted to simplify the bureaucratic processes that this increase would undoubtedly lead to by allowing much work of a minor nature and/or to service installations to be certified as complying by a suitably qualified person (e.g. one who belongs to a particular trade body, professional institution or other approved body).

1.4.4 The future of building control in England and Wales

This section derives its title from a government White Paper (Cmnd 8179) published in February 1981. In paragraph 2 of this document the Secretary of State set out the criteria which any new building control arrangements would be required to satisfy. These were:

- maximum self-regulation
- minimum government interference
- total self-financing
- simplicity in operation.

One out of four (total self-financing) may not seem to be a particularly good result and it has often been the case that the average local authority Building Control Officer has been inadequately prepared through inappropriate education and training to take on the task of assessing compliance with many of the regulation changes listed above. It has been claimed that this problem has been solved by the introduction of Approved Inspectors onto the building control scene. Since these are staffed almost entirely by ex-local authority Building Control Officers it would seem that the net result of the partial privatization of building control has only been to redistribute a finite number of similar people without any improvement in education or training, although the adoption of a more commercial attitude by Approved Inspectors may be a good or bad thing depending on your point of view.

It seems almost inevitable (without a change of government or in government thinking) that the areas of control will increase and that more 'suitably qualified people' will be entitled to certify work as complying with the regulations. It is also likely that local authorities will remain as the final arbiter in matters of enforcement although it likely that their direct involvement in day-to-day building control matters will diminish, to be taken over by the private sector. Indeed, most building control work on new housing is already dealt with by the private sector.

Although the broad subject area covered by the Building Regulations is roughly the same across the European Union (and in former British colonies such as Canada, Australia and New Zealand) the main difference between the system in England and Wales and that in other countries lies in the administrative processes designed to ensure compliance. Our mix of control mechanisms encompassing both public (local authority) and private (Approved Inspector) building control bodies offers choice but also potential conflict. The system is further complicated by the existence of certain 'self-certification' schemes for the installation of, for example, replacement windows and doors or combustion appliances, and some work which has to comply with the regulations but is 'non-notifiable' if carried out by a suitably qualified person.

In fact, we are the only country in the EU with such a 'mixed economy'. Most countries (Scotland and Northern Ireland, Denmark, The Netherlands, the Irish Republic, etc.) use a system run exclusively by the local authority. In Sweden the building control system was privatized in 1995 so that the work of plan checking and site inspections is carried out by a suitably qualified 'quality control supervisor' employed by the building owner, although the local authority still has to be satisfied that the work is being properly supervised and may carry out spot checks and inspections to confirm this.

Some years ago the UK government consulted on proposals to extend control of work governed by the Building Regulations to a range of bodies (some of which could be engaged in design and construction) provided that they were suitably qualified and insured. This would mean, for example, that a firm of architects would be able to take complete control of their own building control processes for work that they had designed without using a local authority or Approved Inspector. Such a market-led system would seem to be in accordance with the all the aims listed at the beginning of this section and, provided that the necessary safeguards could be put in place to prevent corruption and to build public confidence, it would seem to be a sensible way forward. The consultation exercise did not result in any companies being approved to control their own work, although the current system of self-certification of compliance by suitably qualified persons did come out of the exercise. Whether this was caused by political interference, objections from the building control establishment or lack of confidence by companies who still wanted the comfort of a third party to do their regulation checking for them, is not known.

2

Introduction to Part C

2.1 Introduction

The current (2004) edition of Approved Document C became operable on 1 December 2004 with the coming into force of the Building (Amendment) Regulations 2004 (SI 2004/1465). It replaced the 1992 edition (which related to the Building Regulations 1991). Since 1992 there have been several changes relevant to the subject matter covered by Part C. British Standards are being replaced by European ones, the number of regions needing protection from radon has increased, and increased levels of insulation mean that greater consideration needs to be given to condensation risks. The most significant changes concerned contaminated land due to new regulations made under Part IIA of the Environmental Protection Act 1990. This in turn meant that the relevant technical guidance for Building Regulations needed to be revised.

Until the 2004 amendments there were four Requirements under Part C of Schedule 1 to the Building Regulations:

- C1: Preparation of site
- C2: Dangerous and offensive substances
- C3: Subsoil drainage
- C4: Resistance to weather and ground moisture.

During the review and consultation process on the amendment of Part C it was proposed that the principal requirements should be reduced from four to two, namely:

- C1: Preparation of site and resistance to contaminants
- C2: Resistance to moisture.

The rationale for this was that the layout of the 1992 edition of Approved Document C did not logically follow the construction process, since site preparation and resistance to contaminants is concerned with the activities that involve modifying the ground to allow the building to be erected, whereas resistance to moisture deals with all parts of the building envelope in order to exclude the weather and dampness and to create an environment where condensation is less likely to occur.

The following is a summary of the main changes that were brought about by the publication of the 2004 edition of Approved Document C:

- **C1 Site preparation and resistance to contaminants**
  - **Site preparation** – Site investigation is now recommended as the method for determining how much unsuitable material should be removed.
  - **Resistance to contaminants**
    - Requirement C1(2) (see section 2.3 below) now applies to the material change of use of a building (material change of use is fully described in Chapter 4 of the accompanying book in this series *Using the Building Regulations – Administrative Procedures*, first edition 2005, (published by Elsevier Butterworth Heinemann ISBN 0 7506 6257 3)).
    - Remedial measures for dealing with land affected by contaminants have been expanded to include biological, chemical and physical treatment processes.
    - The area of land that is subject to measures to deal with contaminants now includes the land around the building.
    - Guidance on protection from radon is expanded to include buildings other than dwellings.
  - **Sub-soil drainage** – Guidance is included relating to sub-soil drainage and the risk of transportation of waterborne contaminants.
- **C2 Resistance to moisture**
  - New guidance is included on reducing the condensation risk to floors, walls and roofs.
  - Guidance is now provided on the use of moisture resistant boards for the flooring in bathrooms, kitchens and other places where water may be spilled from sanitary fittings or fixed appliances.
  - Additional references are given for assessing the suitability of cavity walls for filling.
  - Check rebates are now recommended in the most exposed parts of the country at the interface between walls and doors and windows.
  - New guidance points to the need for better attention to detail in exposed areas where level access is provided to support Part M, *Access to and*

use of buildings, to ensure adequate provision is made for resistance to moisture.

  ○ Former Requirement F2: *Condensation in roofs*, has been transferred to Part C as it deals with effects on the building fabric rather than ventilation for the health of occupants.

2.2 The 2002 review of Part C and Approved Document C

The Office of the Deputy Prime Minister (ODPM) initiated a review of Part C and Approved Document C in December 2002. In all, 329 bodies deemed to be representative of the issues to be addressed by Part C were consulted, with replies being required to be returned by 10 March 2003. In the event, a considerable amount of interest was shown in the consultation exercise and it was not until 1 December 2004 that Part C came into force. The review considered the impact of the following key issues on Part C:

- the affects of climate change
- moisture ingress
- flooding
- land affected by contaminants
- radon.

2.2.1 The affects of climate change

The review of Part C, referred to above, commenced with a study of what might need to be addressed in Part C should the UK Climate Impacts Programme (UKCIP) 1998 scenarios come to pass. It was concluded that the most likely effects with respect to Part C are:

- increased risk of flooding
- increased risk of summer drought (which in turn may cause shrinkage or cracks in buildings)
- greater risk of rain penetration as a result of increased driving rain
- possible mobilization of ground contaminants (due to increased surface run-off or raised groundwater levels).

New scenarios were published by UKCIP in April 2002 resulting in a review of the 1998 findings. If the 2002 scenarios come to pass there are likely to be more hot summers, winter precipitation will increase and there will be more very wet winters. The greatest effects of this are likely to arise from more frequent flooding and increased driving rain, hence the need to improve and clarify the guidance in these areas in Approved Document C.

2.2.2 Moisture ingress

Under the 1992 edition of Approved Document C the sections that dealt with moisture were oriented towards external sources, e.g. the ground, driving rain, etc. Interstitial condensation from the water vapour produced within buildings has become more important as insulation levels rise and novel constructions with impermeable claddings are introduced, and it was felt that this area of moisture production was dealt with inadequately. Also, the rise in external vapour pressure expected under climate change will reduce the effectiveness of ventilation.

Additionally, since the publication of the 1992 edition of Approved Document C changes to Approved Document M (Access to and use of buildings) have included the need for level access to new dwellings (including guidance on the design of level thresholds). This, of course, has implications for the design both of floors and walls since it is important that water does not penetrate the building fabric at door openings as it could, in turn, cause dampness.

2.2.3 Flooding

The widespread flooding in England and Wales in the autumn of 2000, with severe property damage, focused attention on the various associated issues. These include:

- siting of buildings to avoid flood risk
- protection of buildings by embankments, etc.
- appropriate design and use of materials to minimize damage from flooding
- methods of rapid drying and reinstatement of flooded buildings.

Some of these issues, especially guidance on siting buildings to prevent flooding and appropriate embankments are covered in planning guidance (e.g. Planning Policy Guidance Note PPG 25 *Development and flood risk*, DTLR 2001, see section 5.1 below) and Environment Agency publications, and so are not pertinent to Building Regulations. However, there is still a need to consider appropriate construction techniques for use in areas with a high flood risk. It should be noted that Building Regulations can only address health, safety, welfare and convenience of people. They cannot be made for property protection. Further references on buildings subjected to contaminant-affected floodwater may be found in section 4.4 below.

2.2.4 Land affected by contaminants

The regulatory and technical frameworks dealing with land affected by contaminants have changed significantly since the publication of the 1992 edition of Part C.

New planning advice *Development of land affected by contamination* is currently out for consultation. In due course it will replace the existing guidance on contaminated land in Planning Policy Guidance Note PPG 23 *Planning and pollution control*, ODPM 1997, and will apply in England (PPG 23 is outlined in section 2.5 below). This planning advice will complement the contaminated land regime introduced under Part IIA of the Environmental Protection Act 1990 (EPA), which came into force in April 2000. The regime provides a framework to identify and remove unacceptable risks due to contamination and to bring damaged land back into beneficial use. The EPA contaminated land regime applies to existing land use whereas much of the remediation of land affected by contamination (even that identified under the EPA) will be through the planning system. This reflects the fact that the best means of paying for remediation is often by redevelopment. Despite this, the raft of policy and technical guidance that supports the EPA regime needs to be acknowledged not only in planning advice, but also in Building Regulations. Accordingly, the guidance in support of Part C has been extended to take account of this.

In order to meet future housing needs and to alleviate the pressure on greenfield sites the government's planning policy guidance is proposing that 60% of the required additional housing be built on land that has seen a previous use, and this includes sites that have been subject to a contaminative use (see Planning Policy Guidance Note PPG 3 *Housing*, March 2000).

2.2.5 Radon

At the time the last edition of Part C was published (1992) radon protective measures were only required in Cornwall, Devon and parts of Derbyshire, Northamptonshire and Somerset. This reflected the findings of surveys of radon measurements in existing dwellings undertaken by the National Radiological Protection Board (NRPB) at that time. Since 1992 the NRPB has carried out further surveys including completing national surveys of England and Wales (see NRPB *Radon in dwellings in England: 1997 review*, NRPB-R293, 1997;

NRPB *Radon atlas of England*, NRPB-R290, 1996; NRPB *Radon in dwellings in Wales: Atlas and 1998 review*, NRPB-R303, 1998). Results of these studies were published between 1996 and 1998, and the resulting maps show that most counties in England and Wales have some areas that can be described as 'Radon Affected Areas'.

The new areas identified by the NRPB's findings are used to delineate places where radon protection is required in new dwellings. This was published in BR 211, *Radon: guidance on protective for new dwellings* (see section 4.5.1 below). This publication includes both technical guidance and maps showing where protection is required. The amended guidance was not a result of an amended requirement it was simply bringing the guidance into line with current practice.

2.3 The requirements of Part C

The requirements of Part C are as follows:

- **C1 Preparation of site and resistance to contaminants**
  (1) The ground to be covered by the building shall be reasonably free from any material that might damage the building or affect its stability, including vegetable matter, topsoil and pre-existing foundations.
  (2) Reasonable precautions shall be taken to avoid danger to health and safety caused by contaminants on or in the ground covered, or to be covered by the building and any land associated with the building.
  (3) Adequate sub-soil drainage shall be provided if it is needed to avoid:
      (a) the passage of ground moisture to the interior of the building
      (b) damage to the building, including damage through the transport of waterborne contaminants to the foundations of the building.
  (4) For the purposes of this requirement, 'contaminant' means any substance, which is or may become harmful to any persons or buildings including substances, which are corrosive, explosive, flammable, radioactive or toxic.
- **C2 Resistance to moisture**
  The walls, floors and roof of the building shall adequately protect the building and people who use the building from harmful effects caused by:

    (a) ground moisture
    (b) precipitation including wind-driven spray
    (c) interstitial and surface condensation
    (d) spillage of water from or associated with sanitary fittings or fixed appliances.

2.4 Ways of satisfying the Regulation requirements

2.4.1 Approved Documents and other sources of guidance

Under Section 6 of the Building Act 1984 the Secretary of State may approve any document for the purpose of providing practical guidance with respect to the requirements of the Regulations. Such documents (known as Approved Documents) are intended to be written and illustrated in comparatively straightforward technical terms and are updated at intervals after extensive consultation with a wide range of interested parties who are deemed to be representative of the particular issues covered by the Approved Document under preparation or revision. Part C of Schedule 1 to the *Building Regulations 2000* is supported by Approved Document C entitled 'Site preparation and resistance to contaminants and moisture'. The most recent edition of Approved Document C (2004) replaced the former 1992 edition.

Unfortunately, the Building Regulations and their associated Approved Documents have been getting more complex with every update, often requiring the services of specialist professionals (services engineers, fire engineers, specialists in soil engineering, etc) to make sense of the provisions.

New areas of control are being introduced each year (a proposed Part Q covering electronic communications services is currently under discussion) and the scope of the existing regulations is being extended with each revision.

Furthermore, the current Approved Documents are principally of use in the design of extremely simple and straightforward buildings using mainly traditional techniques. For larger and more complex buildings it is usually better (and more efficient in terms of building design) to use other sources of guidance (British and European Standards, Building Research Establishment Reports, etc.) and although these other source documents are referenced in the Approved Documents no details of their contents or advantages of use are given. As an example of this, the 2004 edition of Approved Document C contains numerous footnote references on virtually every page to no fewer than 109 other sources of guidance.

Additionally, it is often the case that the current Approved Documents fail to provide sufficient guidance just when it is needed, i.e. when it is proposed to deviate from the simple solutions or attempt to design something slightly unusual. Arguably, this has the tendency to encourage adherence to traditional and (perhaps) unimaginative designs and details, and can actively discourage innovation in many modern building designs.

This book (and the others in this series) has been specifically developed to improve on the current situation by addressing different parts of the Building Regulations in separate volumes, thus enabling each Part to be explored in detail.

Each volume expands on the Approved Document guidance by not only describing the direct guidance given, but also by making extensive reference to the other sources of guidance contained in them. These 'alternative approaches' (as they are often described in the Approved Documents) are summarized and the most critical parts of them are presented in the text with indications as to their relative efficacy. Additionally, these guidance documents vary greatly in complexity. At one end of the spectrum some references are clearly designed for use by specialists, such as soils or structural engineers (and involve an advanced level of knowledge of physics, chemistry and mathematics), whereas at the other end there are references pitched at do-it-yourself enthusiasts! In reviewing the references in Approved Document C we have tried to give an indication as to who they would benefit (and whether or not they would be worth acquiring). This information is summarized in the Appendix.

2.4.2 Legal status of Approved Documents

There is no legal obligation to use the Approved Documents when carrying out a building project to which the regulations apply. The applicant is perfectly entitled to choose whether or not to use all or some of the relevant Approved Documents or, indeed, some parts of them. However, care should be taken in certain circumstances not to mix up Approved Document guidance with that from other guidance sources referred to in the specific Approved Document.

Therefore, the legal obligation is to meet the requirements of the Regulations, not the Approved Document guidance. This leaves the designer/developer with the option of devising his or her own solution or of following other authoritative guidance. There may even be circumstances in which it would be unnecessary or inappropriate to follow the guidance in the Approved Document in full, or the Approved Document may offer no guidance on a particular situation being faced.

If adherence to the Approved Document guidance is not mandatory, to what extent do they provide the designer with the confidence that he or she is indeed satisfying the requirements of the regulations and what, if any, is the nature of their legal status? Section 7 of the Building Act 1984 provides that if the guidance in the Approved Documents is followed and it is alleged in any proceedings that the Regulations have been contravened – *'proof of compliance with such a document may be relied on as tending to negative liability'*. Conversely, if the guidance in the Approved Documents has not been followed and it is alleged that the Regulations have been contravened – *'a failure to comply with a document that at that time was approved for the purposes of that provision may be relied upon as tending to establish liability'*. Put more simply, the onus will be on the designer/developer to establish that he or she has met the requirements in some other way and he or she may be called upon by the building control body to demonstrate this by other means. It should be noted that proof of non-compliance with an Approved Document does not necessarily mean that the regulation has been contravened.

A full discussion of ways in which the Regulations may be satisfied is given in Chapter 7 of the accompanying book in this series – *Using the Building Regulations – Administrative Procedures*, published by Elsevier Butterworth Heinemann ISBN 0 7506 6257 3.

2.5 The Environmental Protection Act and other legislation affecting development of contaminated land

The guidance given on resistance to contaminants in Chapter 4 is for the purposes of the Building Regulations and their associated requirements. It should be noted that there may be further provisions for dealing with contaminants contained in planning guidance or legislation made under the regime set out in Part IIA of the *Environmental Protection Act 1990* (EPA) which may be supplementary to the requirements of the Building Regulations. Part IIA (Sections 78A–78YC) was inserted into the EPA by virtue of the *Environment Act 1995* and came into force on 21 September 1995. Section 78B places a duty on local authorities (as defined in Section 78A) to inspect their area from time to time for the purpose of identifying any contaminated land. Local authorities are also required (in collaboration with the Environment Agency) to identify 'special sites' where the contamination is particularly serious and these will then generally be dealt with by the Agency instead of the local authority. Disputes between the local authority and the Environment Agency over the identification of a special site have to be referred to the Secretary of State for a decision. When land has been identified as contaminated, the controlling authority is required to issue a 'remediation notice' on the appropriate person. This will specify the measures which must be taken to remediate the land and will give a timescale. There are consultation measures built in to the procedure and there are rights of appeal in cases of grievance.

The Contaminated Land (England) Regulations 2000 make detailed provisions of a procedural nature to help give full effect to the Part IIA regime, and the statutory guidance provides a basis for enforcing authorities to apply the regime.

Where contaminants are removed, treated or contained as part of the construction works, waste management law may apply. If waste is removed for off-site disposal, the 'Duty of Care' and/or special waste requirements will apply. This will include compliance with the *Environmental Protection (Duty of Care) Regulations 1991*. These regulations impose requirements under Section 34(5) of the EPA on any person who is subject to the duty of care as respects the making and retention of documents and the furnishing of copies of them.

Regulation 2 requires the person transferring (transferor) any controlled waste and the person receiving that waste (transferee) to complete and sign a transfer note at the same time as the written description of the waste is transferred. The transfer note must contain certain information such as:

- the identity of the waste in question
- its quantity
- how it is stored
- the time and place of transfer
- the name and address of the transferor and the transferee
- whether the transferor is the producer or importer of the waste.

One of the most effective means of remediating land affected by contaminants is by way of redevelopment. The redevelopment process is subject to controls under the Town and Country Planning Acts, and Local Planning Authorities follow the guidance in Planning Policy Guidance Note PPG 23 *Planning and pollution control*, ODPM, 1997. This 67-page PPG gives guidance on the relevance of pollution controls to the exercise of planning functions and gives advice on the relationship between authorities' planning responsibilities and the separate statutory responsibilities exercised by local authorities and other pollution control bodies, principally under the Environmental Protection Act 1990 and the Water Resources Act 1991. In particular, it advises that local planning authorities should not seek to duplicate controls which are the statutory responsibility of other bodies (including local authorities in their non-planning functions). It makes clear that close co-ordination among all concerned will be needed to ensure speedy decisions in a complex network of essential regulation. It also provides guidance to planning authorities on the implementation of the European Commission (EC) Waste Framework Directive.

Although environmental protection, planning and Building Regulations have different purposes, their aims are similar. Consequently, the processes for assessing the effects of pollutants and contaminants are similar. An investigation or assessment to determine the characteristics of a site (see Chapter 3, section 3.1 and Chapter 4, section 4.2) can be further developed for Building Regulations purposes when the form and construction of the buildings are known. If appropriate data are gathered at the early stages it should not be necessary to completely re-evaluate a site for Building Regulations purposes.

2.5.1 Consultation with regulatory authorities

During the processes of site investigation, risk assessment and remediation it is quite possible that unexpected events may occur or there may be a change in the expected outcomes. In these case there are likely to be other regulatory authorities which may have an interest in the land affected by the contamination.

The following examples illustrate some likely scenarios:

- The district council should be informed if contaminants are found on a site where the presence of contamination has not been formally recognized through the planning process (also contaminants from the site may be affecting other land or contaminants may be reaching the site from neighbouring land).
- The contamination identified may differ from that which has been previously discussed and agreed with the local planning authority (LPA) or Environmental Health department. This may result in the need for additional discussions with these bodies.
- When a contaminated site is redeveloped, all land quality issues will need to be set out in documents in support of planning approval sent to the local planning authority. Inevitably, the designs will be refined and this may impact on the original risk assessment and remediation strategy. The local planning authority may need to be informed if significant changes are apparent.
- If issues arise that are relevant to the local Environment Agency it may be necessary to consult them. These include:
  - specific duties relating to waste management and the protection of water quality and resources
  - where there is a potential impact on controlled waters if the site is designated as a Special Site under Part IIA of the Environmental Protection Act 1991 (this may mean that an authorization is required, or specific hazards may have been found)
  - remedial measures that may themselves require prior authorization from the Environment Agency (such as abstraction licensing for groundwater treatment and waste management licensing for a number of activities involving contaminated soils).
- The hazards of working on contaminated land should be recognized. This will involve a risk assessment to meet the requirements of the Construction (Design and Management) Regulations 1994 and working procedures will need to be in accordance with the Construction (Health, Safety and Welfare) Regulations 1996. This may necessitate giving notice to the Health and Safety Executive prior to starting work.

2.6 Work on historic buildings

Because the Building Regulations apply not only to new buildings and extensions but also to material changes of use or alterations to existing buildings, a building project may include work on an historic building. Historic buildings include:

- listed buildings
- buildings situated in conservation areas
- buildings which are of architectural and historical interest and which are referred to as a material consideration in a local authority's development plan
- buildings of architectural and historical interest within national parks, areas of outstanding natural beauty and world heritage sites.

Approved Document C recognizes the need to conserve the special characteristics of such historic buildings and contains some guidance as outlined below. Additionally, reference is made to BS 7913: 1998, *Guide to the principles of conservation of historic buildings* (32 pages). This Standard provides guidance (in Clause 7) on the principles that should be applied when proposing work on historic buildings. The following topics are discussed:

- The principle of minimum intervention.
- The knowledge, experience and skill required of owners and occupiers and of those who carry out maintenance and alterations.
- Systematic care including – routine maintenance and housekeeping, protection against fire, protection against other disasters, the need to keep conservation manuals and log books, the need to review the condition of the building every five years and the need to plan for repairs and other work.
- Repair and restoration work including – having a conservative approach to repair, the presumption against and the case for restoration, controls and records in restoration work, materials and details in repair, and restoration work, systems of construction and salvage, reuse and recycling.
- New works including – alterations, extensions and new buildings, the juxtaposition of old and new work, criteria for alteration work, criteria for additions, historic settlements and conservation areas, criteria for new buildings in historic settings.
- The preparation and management of repair and other work including – inspection, survey, research and investigation, approvals and consents, costs, funding and feasibility, specification and preparation of contracts, the use of preliminary contracts, administration and overseeing of repair contracts.

- Research, investigation and recording.
- Interiors, fittings and associated contents.

The guide is intended to provide building owners, managers, archaeologists, architects, engineers, surveyors, contractors, conservators and local authority building control officers with general background information on the principles of the conservation of historic buildings, when considering conservation policy, strategy and procedures.

Therefore, when work is carried out on a historic building, the aim should be to improve resistance to contaminants and moisture where it is practically possible to do so, without prejudicing the character of the historic building or increasing the risk of long-term deterioration of the building fabric or fittings. The local authority's conservation officer should always be consulted when trying to arrive at an appropriate balance between improving resistance to contaminants and moisture, and historic building conservation.

It is recognized in Approved Document C that some particular types of work in historic buildings may warrant sympathetic treatment (perhaps benefiting from advice from conservation professionals). Examples include:

- the need to avoid excessively intrusive gas protective measures
- ensuring that moisture ingress to the roof structure is limited and the roof can breathe.

In the latter case Approved Document C recommends the use of SPAB Information Sheet 4 *The need for old buildings to breathe*, 1986 (4 pages). The aim of this brief information sheet is to examine the differences between traditional and modern materials when used for maintenance purposes and to draw conclusions about the way old buildings should be treated. The information sheet provides guidance on understanding the behaviour of old walls and the problems caused by the use of impervious materials. The differences between traditional and modern materials used for paint systems, external renders, pointing and internal plaster are considered in turn and the paper concludes with a discussion of the correct treatment of old buildings. This paper is essential reading for all concerned with the repair, maintenance and upkeep of old buildings.

Other issues affecting roofs in historic buildings include:

- the sealing of existing service penetrations in a ceiling and the provision of draught proofing to any loft hatches where it is not possible to provide dedicated ventilation to pitched roofs

- keeping new loft insulation clear of the eaves so that any adventitious ventilation is not reduced.

With regard to the need to avoid excessively intrusive gas protective measures, Approved Document C contains the following tips:

- Reduce the rate at which gas seeps into buildings, mainly through floors, by using edge located sumps or sub-floor vents since these are less intrusive than internal sumps or ducts that may involve taking up floors.
- When taking up flagged floors index the stones and record their layout so as to facilitate relaying when work is completed. See also BRE Report BR 267 *Major alterations and conversions: a BRE guide to radon remedial measures in existing dwellings*, 1994 (15 pages). This report is one of a series giving practical advice on methods of reducing radon levels in existing dwellings. The report offers advice on radon-protective measures that can be taken during the planning and implementation of major alteration or conversion works to a building in radon-affected areas. It describes how certain precautionary measures can help to reduce indoor radon levels, and can make it easier to resolve any future radon problem. This is particularly relevant when converting redundant farm or other outbuildings into living accommodation, or carrying out major works such as floor replacement in older properties.
- Disperse radon by using ventilation strategies such as positive pressurization. These systems can often be accommodated in an unobtrusive manner.
- If internal mechanical ventilation is used to disperse ground gases, it may affect the functioning of combustion appliances and may lead to the spillage of products of combustion into the building. Guidance on this can be found in Good Building Guide 25 *Buildings and radon* (see section 4.5.1 below).

PART 2

Site preparation and resistance to contaminants

Clearance or treatment of unsuitable material

3.1 Preparation of site – site investigation

In order to prepare a site for construction work it will usually be necessary to carry out a site investigation. This will entail a study of site conditions to determine their probable influence on the design, construction and subsequent performance of a building. The following items may be encountered and are discussed in Approved Document C:

- unsuitable material such as turf and roots
- mature trees which might affect services, floor slabs, oversite concrete and foundations
- pre-existing foundations, services and other infrastructure, and buried tanks
- fill or made up ground
- contaminants
- high water table, which might necessitate sub-soil drainage to avoid damage to the building.

Additionally, the site investigation will identify the type and nature of the sub-soil and its load-bearing capacity which will be of use in the design of foundations in accordance with Part A, Structure of Schedule 1 to the *Building Regulations 2000* (a book in this series is in preparation for Part A).

The site investigation will normally consist of the following stages:

(1) **Planning stage.** In this stage the objectives, scope and requirements of the investigation are set so as to enable it to be planned and carried out efficiently and so that the required information may be provided.

(2) **Desk study.** A desk study is a review of the historical, geological and environmental information about the site. The information can be obtained much more quickly and cheaply than information from boreholes and trial pits and a great deal of factual information is publicly available in the UK from sources such as geological maps, Ordnance Survey sheets, aerial photographs, computer satellite databases, geological books and records, local authority building control data, mining records, reports of previous site investigations etc.

(3) **Site reconnaissance or walk-over survey.** The site reconnaissance or walk-over survey aids the design of the main investigation in that it identifies actual and potential physical hazards. The object of the walk-over survey is to check and make additions to the information already collected during the desk study. The site and its surrounding area are visited and covered carefully on foot. When carrying out a walk-over survey it is also possible to gather information from local authorities, local inhabitants and people working in the area, such as builders, electricity and gas workers. On completion of the survey a structured report can be produced from the information gathered at the site and from the local enquiries.

(4) **Main investigation and reporting.** This will usually include an examination of the geotechnical properties of the ground and should be designed to verify and expand information previously collected from the desk study and site walk-over survey. The investigation will usually include intrusive and non-intrusive sampling and testing by means of trial pits and boreholes to provide soil parameters for design and construction.

In determining the extent and level of investigation it will be necessary to consider the type of development proposed and the previous use of the land. Typically the site investigation should include:

- susceptibility to groundwater levels and flow
- underlying geology
- ground and hydro-geological properties.

A geotechnical site investigation should:

- identify physical hazards for site development
- determine an appropriate design
- provide soil parameters for design and construction.

Where there is concern that the site might be affected by contaminants, a combined geotechnical and geo-environmental investigation should be considered. Section 2 of Approved Document C *Resistance to contaminants* (which is considered in

Chapter 4) contains guidance on assessing and remediating sites affected by contaminants.

3.1.1 Site investigation – sources of guidance

A number of guidance sources are given in Approved Document C. Comprehensive guidance on site investigations in general may be found in British Standard 5930: 1999 *Code of practice for site investigations*, whereas for low-rise buildings the following documents may also be consulted:

- BRE Digest 322 *Site investigation for low-rise building: procurement*, 1987
- BRE Digest 318 *Site investigation for low-rise building: desk studies*, 1987
- BRE Digest 348 *Site investigation for low-rise building: the walk-over survey*, 1989
- BRE Digest 381 *Site investigation for low-rise building: trial pits*, 1993
- BRE Digest 383 *Site investigation for low-rise building: soil description*, 1993
- BRE Digest 411 *Site investigation for low-rise building: direct investigations*, 1995
- BS 8103: Part 1: 1995 *Structural design for low rise buildings*.

British Standard 5930: 1999 Code of practice for site investigations
BS 5930 is a document of over 200 pages dealing with the investigation of sites for the purposes of:

- assessing their suitability for the construction of civil engineering and building works
- gaining knowledge of the characteristics of a site that affect the design and construction of civil engineering and building works and the security of neighbouring land and property.

The expression 'site investigation' is used in its wider sense in the Code where it encompasses more than just the exploration of ground. BS 5930 uses the term 'ground investigation' to cover this but recognizes that the treatment of ground investigation is always likely to be more detailed than the treatment of other aspects of site investigation.

The Code is divided into the following sections:

- *Section 1: Preliminary considerations* – deals with technical, legal or environmental factors that are usually taken into account when selecting a site (or in determining the suitability of a selected site) and in the preparation of the design of the works.

- *Section 2: Ground investigations* – discusses general aspects and planning of ground investigations. The objectives of ground investigations are to:
  - ○ obtain reliable information to produce an economic and safe design
  - ○ assess any hazards (physical or chemical) associated with the ground
  - ○ meet tender and construction requirements.

  The investigation should be designed to verify and expand information previously collected.
- *Section 3: Field investigations* – describes methods of ground investigation including excavation, boring, sampling, probing and tests in boreholes.
- *Section 4: Field tests.*
- *Section 5: Laboratory tests on samples.*
- *Section 6: Description of soils and rocks* – deals with the terminology and systems recommended for use in describing and classifying soil and rock materials and rock masses.
- *Section 7: Reports and interpretation* – deals with the preparation of field reports and final borehole logs, the interpretation of the data obtained from the investigation and the preparation of the final report.

The Code is a key document for practitioners in the field of site investigation. Other users, such as designers, building control staff and general builders may also find it useful in giving a general understanding of the scope of site investigation and the techniques of ground investigation. For these purposes, Sections 1 and 2 will found most useful since they augment the limited information given in Approved Document C and described in section 3.1 above.

BRE Digests – series covering site investigation for low-rise building

This series of BRE Digests is specifically aimed at low-rise housing and is written in non-technical language giving practical advice on all issues concerned with site investigation. They are particularly useful for the non-specialist (and for students). The topics discussed include:

- Procurement of the site investigation work (Digest 322, 8 pages) such as the value of site investigation, the steps that should be involved, and the contractual methods that can be employed to engage suitable specialists to carry out the work. The Digest recommends that at least 0.2% of the cost of the project should be spent on site investigation and goes on to describe two systems for carrying out the work as follows:

○ System I deals with the appointment of a geotechnical adviser with the separate employment of a contractor for physical work, testing and reporting as required.

○ System II covers package deal contracts, with desk study, planning and execution of field and laboratory work and reporting being carried out by one company or a consortium.

The Digest goes on to describe the advantages and disadvantages of each procurement system

- Desk studies (Digest 318, 12 pages). These involve the collection of as much information as possible about a site from sources such as geological maps, Ordnance Survey maps, air photographs, geological books, civil engineering magazines, mining records and reports of previous site investigations. This information can be obtained much more cheaply and quickly than that obtained, for example, from boreholes and trial pits. The Digest discusses the influence desk studies can have on the identification of ground problems and contains a very useful checklist covering questions which need to be answered in order to assess which ground problems might occur. These are related to:

○ topography, vegetation and drainage

○ ground conditions

○ the proposed structure.

The Digest is completed with a series of seven case studies illustrating the use of various types of published information obtained during desk studies.

- The walk-over survey (Digest 348, 8 pages). The Digest describes the objectives of the walk-over survey (to check and make additions to the information already collected during the desk study) and shows the importance of visiting the site and its surrounding area and covering it carefully on foot. During this process, questions can be asked of local authorities, local inhabitants and people working in the area, such as builders, electricity and gas workers, in order to obtain the benefit of their local knowledge before the production of a structured report based on the information gathered. The Digest covers the process of carrying out the survey (including a list of basic tools that can be used) and gives examples of features that can be observed and recorded as part of the survey. A list of possible local sources of information is given and brief notes are provided on the following topics:

○ Topography of the site, including slope angles

○ Expected groundwater conditions

○ Geological setting

○ Location, size and species of any trees, shrubs or hedges, either at present or in the past, with notes on dates when removed

- ○ Chemical aggressivity of ground and groundwater
- ○ Probability of pre-existing slope instability
- ○ Position of existing and demolished structures
- ○ Possible extent and dates of mineral extraction and mining in the area
- ○ Evidence for the existence of made ground on site
- ○ Possible locations for structures
- ○ Likely types and loading of structures
- ○ Access for excavators and boring/drilling rigs during ground investigation
- ○ Position of services (e.g. electricity, water, gas, telephone, sewers).
- • Trial pits (Digest 381, 12 pages). Shallow trial pits can provide an economic and versatile way of examining soil conditions *in situ*. The Digest recommends that the investigation, rather than the cost, should be the controlling factor in the selection of the investigatory methods. Trial pits can be dug mechanically or by hand and are used to:
  - ○ facilitate detailed examination of the ground down to a depth of about 4 m
  - ○ obtain undisturbed samples – usually done by driving sample tubes into the base or sides of the trial pit, or by taking high-quality block samples cut from a bench formed in the trial pit
  - ○ for more intensive *in situ* testing, e.g. plate-bearing tests
  - ○ to examine qualities of the ground such as, ease of excavation or subsequent behaviour of excavated ground.

  It may not always be necessary to carry out *in situ* tests in the case of low-rise buildings if accurate soil descriptions can be made. The digest describes the techniques of trial pit excavation and the characteristics of various types of soil when excavated, including methods of support. Since the primary purpose of excavating trial pits is to examine the type and sequence of geological deposits found on the site, an essential part of the examination is a complete description of the soils and rocks present (including fill in made up ground). The digest describes techniques for logging the trial pit information and provides an example of a blank and completed form that can be used for this purpose. Techniques for obtaining undisturbed samples are described as well as backfilling methods. The Digest lists the following important points on a trial pit log:
  - ○ date of excavation
  - ○ location of pit on the site
  - ○ ground level of pit
  - ○ overall dimensions of the pit
  - ○ excavation techniques used
  - ○ ease of excavation
  - ○ groundwater conditions encountered
  - ○ stability of sides of trial pit
  - ○ whether logged from the surface or *in situ*
  - ○ full soil description and stratum depths

- ○ exact location and orientation of undisturbed samples relative to pit floor and sides
  - ○ exact location relative to pit floor and sides of disturbed samples and *in situ* tests
  - ○ *in situ* testing.
- Soil description, (Digest 383, 12 pages). The Digest explains how to make an accurate description of the soil. Although rock descriptions are not considered in any detail many aspects of soil description are also applicable to them. The Digest discusses the two main aspects of the field identification and engineering description of soils. These are:
  - ○ material characteristics which reflect the mineralogy of the soil grains and the range of grain sizes present
  - ○ the *in situ* soil characteristics which are related to such factors as the density of packing of the soil particles and to any discontinuities that may be present in the soil mass (e.g. joints and fissures).
- Other important information concerns:
  - ○ the geological formation
  - ○ the age and composition of a deposit
  - ○ special features that apply such as the presence of voids, tree roots or human-made objects.
- Each homogenous soil layer of a profile should be described systematically using the following terms:
  - ○ soil type
  - ○ moisture condition
  - ○ colour
  - ○ consistency (or strength)
  - ○ structure
  - ○ other special features
  - ○ origin
  - ○ groundwater conditions.

  The Digest describes the characteristics of different soil materials in terms of grain size and shows how this may be related to plasticity. The importance of moisture condition of the soil is discussed and the significance of soil colour is brought out. Other issues covered include soil consistency, strength of granular soils, consistency of clay soils, and strength of silts and peat. Structural features of soil are also discussed in terms of discontinuities and bedding.
- Direct investigations, (Digest 411, 12 pages). Direct investigations are carried out to:
  - ○ check the information from the desk study
  - ○ obtain any additional information required to ensure safe and economic construction.

Advice is given on planning direct investigations which should consist of:

○ identifying the depths of investigation required at different locations around the site (e.g. boreholes should always penetrate completely through made ground or infilling)

○ identifying suitable *in situ* and laboratory testing methods for the expected soil conditions and the parameters required

○ deciding on the number of exploratory holes and the sampling and testing frequency, making allowances for the presence of unforeseen ground or groundwater conditions.

• It will also be necessary to record the basis of the planned site investigation and the expected ground conditions. The specialist contractor who carries out the work will then know if the ground conditions he or she encounters are unforeseen (in which case it may be necessary to alter the scope of the field or testing work).

• This digest repeats, updates and consolidates a great deal of the information provided by those described above and is most useful if a general idea of the scope and nature of site investigation is required.

British Standard 8103: Part 1: 1995 Structural design for low rise buildings: code of practice for stability, site investigation, foundations and ground floor slabs for housing

This 32-page document gives recommendations for the structural design of low-rise housing and covers the stability of the structure, site investigation and foundations and ground floor slabs used in the construction. Foundations comprising strip footings or trench fill founded in normal ground are the only type described. The parts of the Standard concerned with site investigation are contained in Section 6 and provide a brief summary of the information covered by the references mentioned above. On its own, Section 6 is of little use if the other references have been obtained. However, when taken as a whole the Standard provides much useful information on foundations and ground floor slabs and, curiously, information on the tying of timber floors and roof structures to walls.

3.2 Preparation of site – clearance or treatment of unsuitable material

3.2.1 General

The ground to be covered by the building is required to be reasonably free from any material that might damage the building or affect its stability. This includes vegetable matter, topsoil and pre-existing foundations. (See Requirement C1(1).)

Decaying vegetable matter could be a danger to health and it could also cause a building to become unstable if it occurred under foundations. Approved Document C, therefore, recommends that the site should be cleared of all turf and vegetable matter at least to a depth to prevent future growth. This might not apply to buildings used solely for storage of plant or machinery in which the only persons habitually employed are store people, etc. engaged only in taking in, caring for or taking out the goods. Other similar types of buildings where the air is so moisture-laden that any increase would not adversely affect the health of the occupants are also excluded.

Below-ground services (such as foul or surface water drainage) should be designed to resist the effects of tree roots. This can be achieved by making services sufficiently robust or flexible and with joints that cannot be penetrated by roots. Consideration should be given to the removal of roots where they could pose a hazard to below-ground services.

Where a site has been previously built on it will be necessary to consider if it contains any pre-existing foundations, services, buried tanks and any other infra-structure etc. that could be a danger to persons in and about the building and any land associated with the building. In this context, Approved Document C defines *building and land associated with the building* as 'the building and all the land forming the site subject to building operations which includes land under the building and the land around it which may have an effect on the building or its users'. This definition is further clarified in paragraph 2.11 of Approved Document C where it refers to the 'area of the site subject to building operations', i.e. 'those parts of the land associated with the building that include the building itself, gardens and other places on the site that are accessible to users of the building and those in and about the building'.

3.2.2 Shrinkable clay soils

Where shrinkable clays soils are present on a site, the presence of mature trees can exacerbate the tendency of the soil to cause heave and subsidence and this may lead to damage to services, floor slabs and oversite concrete. On such soils, the potential for damage should be assessed and Approved Document C gives guidance, in general terms, on the likely potential for volume change for some commonly occurring clays. Reference should be made to Diagram 1 from Approved Document C (reproduced here in Figure 3.1) to ascertain the type of clay that might be occurring. When used with Table 1 from Approved Document C (shown in Table 3.1) the volume change potential can be approximately assessed.

For more detailed guidance, reference may also be made to BRE Digest 298 *Low-rise building foundations: the influence of trees in clay soils*, 1999. In shrinkable

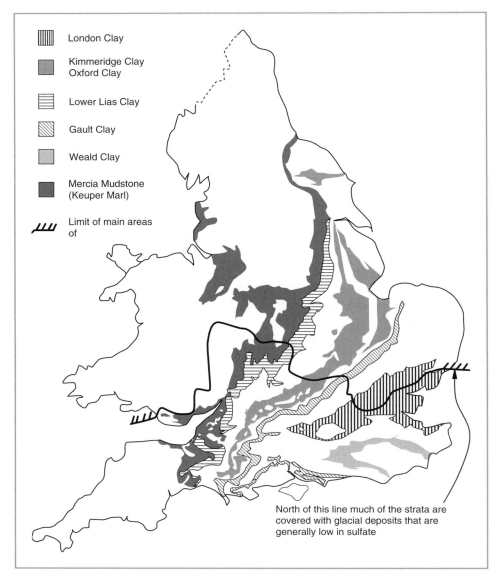

Figure 3.1 Approved Document C – Diagram 1: Distribution of shrinkable clays and principal sulfate/sulfide-bearing strata in England and Wales

clay soils, soil shrinkage caused by the removal of water by trees can lead to foundation subsidence. Conversely, soil swelling can occur following tree removal as the soil moisture levels recover. This can lead to foundation heave. This 8-page Digest gives simple guidance on minimizing these effects in clay soils and discusses some dangers in current foundation practice. It is highly recommended for anyone concerned with the design and construction of low-rise buildings on shrinkable

Table 3.1 Approved Document C, Table 1 – Volume change potential for some common clays

Clay type	Volume change potential
Glacial Till	Low
London	High to very high
Oxford and Kimmeridge	High
Lower Lias	Medium
Gault	High to very high
Weald	High
Mercian mudstone	Low to medium

clay soils. Topics covered by the Digest include:

- trees, clay, climate and ground movement – a discussion of the mechanisms involved by both direct and indirect action and the types of foundation movement that can occur
- the proximity of trees to buildings – how to establish a 'safe' distance
- trees in relation to new and existing foundations – new tree planting, the presence of existing trees and the removal of trees
- preventing heave damage to trench fill foundations.

Sometimes it becomes necessary to remove significant quantities of soil. Approved Document C makes reference to a number of sources of guidance which are outlined as follows:

- BRE Digest 241 *Low rise buildings on shrinkable clay soils*: Part 2, 1993. This Digest reviews the use of traditional and trench fill strip foundations on sites containing shrinkable clay soils where trees and other vegetation are present. The use of very deep trench fill foundations (up to 3 m) is criticized and advice is given on the use of driven mini-shell pile foundations and short-bored pile foundations as an alternative to traditional strip. The precautions that should be taken when using piled foundations in shrinkable clays are described and a useful comparison is given between the costs of installing piled, trench fill and traditional strip foundations. Surprisingly, piled foundations with reinforced ground beams proved to be the most cost-effective solution.
- BRE Digest 242 *Low rise buildings on shrinkable clay soils*: Part 3, 1993. (Note: Although this document is referred to in Approved Document C, the author was unable to trace it, even after searching the BRE web site.)

- The Foundation for the Built Environment (FBE) report *Subsidence damage to domestic buildings: lessons learned and questions remaining*, Foundation for the Built Environment (FBE), 2000. This 35-page publication, written by R. M. C. Driscoll and M. S. Crilly and published by the BRE Centre for Ground Engineering and Remediation, is the first in a developing series of publications about the wide-ranging research projects commissioned by the FBE to assist those working in the construction and associated sectors. The report contains some useful research evidence on the nature of the problem of subsidence, and reviews the findings and contents of some of the Digests mentioned above. Of particular interest is Chapter 2 'Lessons learned' and Chapter 3 'Questions remaining'. These chapters discuss the fundamental technical issues in terms of the soil, the tree, the weather, the foundations and the building. They consider the factors concerned in dealing with a case of damage by reference to its significance and its causes and the practicalities of diagnosing subsidence. Remedies for subsidence damage are considered as are methods for avoiding it. Other issues covered by the report include the socioeconomic factors, insurance cover, geographical loading, and the position in other countries. Future technical developments are covered in Chapter 4. Overall, the report provides useful background reading on the problems of, and solutions to, subsidence damage to domestic buildings; however, its lack of practical construction examples means that it cannot be used as a substitute for some of the BRE Digests and other more practical references mentioned above.
- The National House Building Council (NHBC) Standards Chapter 4.2 *Building near trees*, 2003 (52 pages). This is one of a range of standards produced by the NHBC for use in connection with their Buildmark 10-year housing warranty. This publication describes the design standard required, restates the statutory requirements and considers the design and construction of foundations suitable for building in shrinkable soils and where trees are adjacent to dwellings. It can also be useful in assessing the effects of remaining trees on services and building movements close to the building. Like all the NHBC Standards this one is easy to read, well illustrated and relatively non-technical. It is aimed at designers and contractors of new housing.

3.2.3 Building on fill

Sites which contain fill or made ground can present particular problems related to the compressibility of the ground and its potential for collapse when wetted. Appropriate remedial measures may need to be taken to prevent differential settlement from causing damage to the building.

Guidance on these issues may be found in BRE Digest 427 *Low-rise buildings on fill* and BRE Report BR 424 *Building fill: Geotechnical aspects*, 2001.

Digest 427 (*Low-rise buildings on fill*) is in three parts.

Part 1 (8 pages) provides a general introduction, describes the causes of settlement of fill and gives guidance on the likely magnitude of settlement in different situations including data on the compressibility of various fill materials. A simple classification of fill materials is based on estimated fill movements.

Part 2 (12 pages) recommends that construction on existing filled areas should be preceded by careful investigation of the site which should be both appropriate and adequate. The site investigation may indicate that ground treatment is required before construction to improve the load-carrying properties of the fill, reducing or forestalling the effects of the unexpected. This Part describes approaches to site investigation, techniques for improving the load-carrying characteristics of non-engineered fills (such as dynamic compaction, rapid impact compaction, vibrated stone and concrete columns) and aspects of foundation design for use on fills.

Part 3 (8 pages) presents guidance for the specification and control of fills which are to be placed so that they can safely support low-rise buildings without the occurrence of damaging movements. Such fills are described as foundation fills. When new fill is to be placed, careful selection of the material and controlled placement should ensure that the fill forms an adequate foundation material. Part 3 is particularly relevant to building developments where relatively small volumes of fill are being placed and, consequently, the resources available for design, testing etc. are quite limited.

Digest 427 gives guidance only of a general nature and it must be recognized that the services of civil, structural and geotechnical engineers will usually be needed.

BRE Report BR 424 *Building fill: Geotechnical aspects* by J. A. Charles and K. S. Watts (208 pages) provides a detailed account of BRE research findings and their significance for appropriate and successful building developments on fill. Part A deals with the engineering behaviour of fills, and Part B examines construction on fills. Brief case histories of field performance are presented in Part C and these mostly describe sites where BRE has made measurements of fill behaviour, but some additional case histories, in which monitoring has been carried out by other parties are included where necessary to give a more complete picture. (Parts A and B make extensive use of these records.) Field monitoring has shown that in most situations the fill settlement that damages buildings has causes other than the weight of the building. This means that the concept of bearing capacity is not adequate to define the load-carrying characteristics of many fills. Settlements caused by other physical factors, and in some cases by chemical or biological processes, need to be assessed. A particular hazard for poorly compacted partially saturated fills is a reduction in volume which can occur when the fill is first inundated with water.

This extremely detailed report will be essential reading for all professionals who are specifically concerned with the appraisal of fill materials on site (especially for the development of 'brownfield' sites) but it does contain a great deal of highly technical ground engineering material which will only be of interest to trained engineers. If information of a general nature is required then BRE Digest 427 is recommended instead.

Resistance to contaminants

4.1 Introduction

Reasonable precautions must be taken to prevent any contaminants found on or in the ground from causing a danger to health and safety (see requirement C1(2)). This is, of course, the ground covered (or to be covered) by the building and includes any land associated with the building (see definition in section 3.2.1 above).

There is a special definition of *contaminant* for the purposes of requirement C1(2) – any substance which is or could become harmful to persons or buildings including substances, which are toxic, corrosive, explosive, flammable or radioactive.

The term '*contaminated land*' is defined in Part IIA of the Environmental Protection Act 1990 as 'any land which appears to the local authority in whose area it is situated to be in such a condition, by reason of substances in, on or under the land, that significant harm is being caused, or there is a significant possibility of such harm being caused, or pollution of controlled water is being or is likely to be caused'. This harm can be to any thing (receptor) that could be harmed by substances in the land, whereas the Building Regulations can only be concerned with the health and safety of persons in or about buildings.

Contaminants which occur on sites can be liquids, solids or gases and can arise out of a previous use of land, especially where this was related to an industrial undertaking. In recent years problems have arisen from the emission of landfill gas from waste disposal sites where it is common for biodegradable waste to be buried. Even sites which have been used for rural purposes, such as agriculture, may be contaminated by pesticides, fertilizer, fuel and oils and decaying matter of biological origin. In 1984 the author remembers encountering an edge-of-town greenfield site contaminated with the carcasses of cattle slaughtered in an outbreak of foot and mouth disease from the 1950s.

4.1.1 Sources of contamination

Where a site is being redeveloped, knowledge of its previous use, from planning or other local records, may indicate a possible source of contamination. Table 2 to Section 2 of Approved Document C (AD C) (reproduced here in Table 4.1) lists a number of site uses that are likely to contain contaminants. It should not be considered to be an exhaustive list. It is derived from the *Industry Profile* guides published by the former Department of the Environment (*Department of the Environment Industry Profiles, 1996*). Each of these profiles considers a different industry which has the potential to cause contamination. The particular contaminant associated with the industry is identified together with details of where it may be found on the site and the routes it might take to migrate to other areas.

Table 4.1 Approved Document C, Table 2 – Examples of sites likely to contain contaminants

Animal and animal products processing works
Asbestos works
Ceramics cement and asphalt manufacturing works
Chemical works
Dockyards and docklands
Engineering works (including aircraft manufacturing, railway engineering works,
  shipyards, electrical and electronic equipment manufacturing works)
Gas works, coal carbonisation plants and ancillary by-product works
Industries making or using wood preservatives
Landfill and other waste disposal sites
Metal mines, smelters, foundries, steel works and metal finishing works
Munitions production and testing sites
Oil storage and distribution sites
Paper and printing works
Power stations
Railway land, especially the larger sidings and depots
Road vehicle fuelling, service and repair: garages and filling stations
Scrap yards
Sewage works, sewage farms and sludge disposal sites
Tanneries
Textile works and dye works

Further information on the assessment of land affected by contaminants is presented in Defra/Environment Agency Contaminated Land Research Report CLR 8 *Potential contaminants for the assessment of land*, 2002 (this is referred to incorrectly in AD C as *Priority* contaminants for the assessment of land). CLR 8 identifies priority contaminants (or families of contaminants), selected on the basis that they are likely to be present on many current or former sites affected by industrial or waste management activity in the United Kingdom in sufficient concentrations to cause harm; and that they pose a risk, either to human health, buildings, water resources or ecosystems. It also indicates which contaminants are likely to be associated with particular industries. The primary purpose of the selection has been to provide the Department for Environment, Food and Rural Affairs (DEFRA) with a guide to the substances it should cover in its research work on contaminated land. CLR 8 deals with the criteria for selection, the risks such contaminants are likely to pose, and the reasoning behind non-selection of certain substances.

CLR 8 is one of a series of reports published by DEFRA and the Environment Agency that is relevant to the assessment of the risks to human health arising from long-term exposure to soil contamination. These reports take into account the wider DEFRA and predecessor department's guidelines on assessing and managing environmental risk (Department of Environment, Transport and the Regions (DETR), Environment Agency and Institution of Environmental Health (IEH), 2000). It is strongly recommended that each report in the series should be read in conjunction with the others, and Table 1.1 in CLR 8 provides more information on the content of each report. These are summarized as follows:

- CLR 7 *Assessment of Risks to Human Health from Land Contamination: An Overview of the Development of Soil Guideline Values and Related Research* (DEFRA and Environment Agency, 2002) (40 pages). CLR 7 serves as an introduction to the other reports in the series. It sets out the legal framework, in particular the statutory definition of contaminated land under Part IIA of the Environmental Protection Act (EPA) 1990, the development and use of Soil Guideline Values, and references to related research.
- CLR 9 *Contaminants in Soil: Collation of Toxicological Data and Intake Values for Humans* (DEFRA and Environment Agency, 2002) (48 pages). This report sets out the approach to the selection of tolerable daily intakes and Index Doses for contaminants to support the derivation of Soil Guideline Values.
- CLR TOX 1–10 (DEFRA and Environment Agency, 2002). These reports detail the derivation of tolerable daily intakes and Index Doses for the following contaminants, which are arsenic, benzo[*a*]pyrene, cadmium, chromium, inorganic cyanide, lead, phenol, nickel, mercury and selenium.

- CLR 10 *The Contaminated Land Exposure Assessment Model (CLEA): Technical Basis and Algorithms* (DEFRA and the Environment Agency, 2002) (162 pages). This report describes the conceptual exposure models for each standard land use that are used to derive the Soil Guideline Values. It sets out the technical basis for modelling exposure and provides a comprehensive reference to all default parameters and algorithms used.
- CLR GV 1–10 (DEFRA and Environment Agency, 2002). These reports set out the derivation of the Soil Guideline Values for the following contaminants, which are arsenic, benzo[*a*]pyrene, cadmium, chromium, cyanide (free, simple, and complex inorganic compounds), lead, phenol, nickel, mercury (inorganic compounds) and selenium.
- CLR 11 *Model Procedures for the Management of Contaminated Land* (DEFRA and the Environment Agency, 2004) (204 pages). This report incorporates existing good technical practice, including the use of risk assessment and risk management techniques, into a systematic process for identifying, making decisions about and taking appropriate action to deal with contamination, in a way that is consistent with UK policy and legislation.

These reports will mainly be of use to regulators, developers and their advisers who will be interested in the selection because it identifies contaminants that are important on a national basis. The selection can also be used for the assessment of individual sites. However, if the selection is put to this use it should be noted that it will not be necessary to investigate all the selected substances on every industrial site, and, depending on their particular industrial histories, some sites might need to be investigated for substances that are not included in the selection.

4.1.2 Naturally occurring contaminants

In certain parts of the country, the following naturally occurring contaminants can arise from the underlying geology:

- mining areas:
  - certain heavy metals, such as cadmium and arsenic
  - gases such as methane and carbon dioxide (mainly from coal mining areas)
- carbon dioxide and methane gases arising from organic rich soils and sediments (such as peat and river silts) (see section 4.6 below).

Further information on these contaminants including their geographical extent, associated hazards, site investigation methods, and protective measures that can be taken can be obtained from the following Environment Agency publications:

- in England – Environment Agency R & D Technical Report P291 Information on land quality in England: Sources of information (including background contaminants)
- in Wales – Environment Agency R & D Technical Report P292 Information on land quality in Wales: Sources of information (including background contaminants).

Environment Agency R & D Technical Report P291 (113 pages) and Environment Agency R & D Technical Report P292 (98 pages) taken together present an overview of information on land quality in England and Wales, carried out for the Environment Agency who are required, under the Environment Act (1995), to form an opinion on the state of pollution of the environment. The main objective of the study was to identify data sets that can be used to assess land contamination. This included a review of existing knowledge on background levels of contaminants in English soils. The research was undertaken by extensive literature review and through consultation with Environment Agency staff and external organizations, in order to identify a wide range of land quality information, including soil survey data, environmental monitoring data, research studies, and land use information. Particular emphasis was placed on identifying data sets representative of land quality at regional to national scales, rather than collating site-specific and local scale information. Results of the study indicated a number of knowledge gaps in land quality information which need to be addressed in order to prioritize research needs and formulate national policies.

These reports will be of interest only to policy makers and developers, when making decisions about the quality of land which it is intended to develop. A number of information sources are identified which may be useful in providing data on particular contaminants at particular sites.

Other naturally occurring contaminants include:

- the radioactive gas radon
- sulphates.

Contamination by radon gas and its products of decay has led to concern over the long-term heath of occupants of affected buildings. Measures to protect buildings and occupants against ingress of radon gas are considered in section 4.5 below. Sulphate attack affects concrete floor slabs and foundations and can cause failure and disruption. Measures can be taken (such as the use of sulphate-resisting cement) in those areas where naturally occurring sulphates are present.

Diagram 1 and Table 1 from Approved Document C (illustrated in Figure 3.1 and Table 3.1 respectively) show the principle areas of sulphate-bearing strata in England and Wales. Reference may also be made to BRE Special Digest SD1 *Concrete in aggressive ground*, 2003, where guidance will be found on investigation, concrete specification and design to mitigate the effects of sulphate attack. This 70-page document aims to provide practical guidance to concrete designers, contractors, specifiers and producers on the specification of concrete to resist chemical attack. It will also be of use to ground specialists in the assessment of ground in respect of aggressiveness to concrete. The Digest is divided into 6 Parts (A to F) as follows:

Part A – introduction, describes the problem of chemical attack and outlines the scope and structure of the remainder of the publication.

Part B – describes modes of chemical attack and considers the mechanisms of the principal types, (such as sulphate and acid attack) and the action of aggressive carbon dioxide.

Part C – covers assessment of the chemical aggressiveness of the ground. It gives procedures for classifying a location in terms of Aggressive Chemical Environment for Concrete Class (ACEC Class) for both natural ground and brownfield sites.

Part D – gives recommendations for the specification of concrete for general cast-in-situ use in the ground. In some cases, where conditions are highly aggressive, additional protective measures (APMs) are recommended.

Part E – gives recommendations for specifying surface-carbonated precast concrete for general use in the ground.

Part F – includes design guides for specification of specific precast concrete products, including pipeline systems, box culverts, and segmental linings for tunnels and shafts. Part F also covers specification of precast concrete masonry units (concrete blocks) for aggressive ground conditions.

4.2 Resistance to contaminants – risk assessment

4.2.1 Introduction and general concepts

In order to develop potentially contaminated land safely it is necessary to carry out a risk assessment. This is normally done by adopting a tiered approach in which an increasing level of detail is required as the tiers are worked through. For a full risk assessment it will be necessary to progress through the following three tiers:

- preliminary risk assessment
- generic quantitative risk assessment (GQRA)
- detailed quantitative risk assessment (DQRA).

The need for a risk assessment will usually be identified during the first stages of the site investigation (desk studies and site walk-over survey). Once this need has been identified, a preliminary risk assessment must always be undertaken. The extent to which it is necessary to do a more detailed risk assessment will depend on the situation and outcome of the preliminary assessment. This may indicate that it will be necessary to do only one or other (or both) of the more detailed risk assessments.

The general approach to risk assessment described above is based on the concept of the relationship between the *source* of contamination (the contaminants found on or in the ground), the *pathway* taken by the contaminants (e.g. ingestion, inhalation, direct contact, attack on building materials and services) and the *receptor* of those contaminants (i.e. buildings, building materials and services and people). This 'source–pathway–receptor' relationship, or pollutant linkage, is illustrated in Figure 4.1.

The development of contaminated land has the inevitable consequence of introducing receptors (buildings, building services, building materials and people) onto the development site. In order to mitigate the effects of the pollutants on the receptors the pollutant linkages must be broken. This can be achieved in a number of ways, for example, by:

- using physical, chemical, biological or other processes to treat the contaminant so as to eliminate or reduce its toxicity or harmful properties
- removing or blocking the contaminant pathway (e.g. by installing barriers to prevent migration or protective layers to isolate the contaminant)
- removing or protecting the receptor (perhaps by the use of appropriately designed building materials, or by changing the form or layout of the development)
- removing the contaminant by excavating the contaminated material.

4.2.2 Risk assessment stages

For each of the tiers mentioned above (preliminary risk assessment, generic quantitative risk assessment (GQRA) and detailed quantitative risk assessment (DQRA)) AD C recommends that the Defra/Environment Agency Contaminated Land Research Report CLR 11 *Handbook of model procedures for the management*

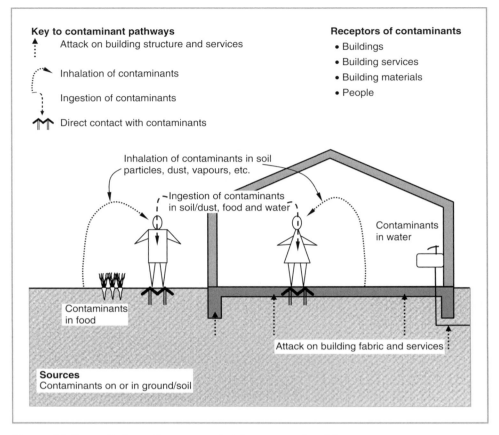

Figure 4.1 Source–Pathway–Receptor – typical site conceptual model

of contaminated land, 2003 should be consulted (see section 4.1.1 above for a summary of this document). At the time that AD C was published this was only a consultation draft. It has since been published (in September 2004) as Contaminated Land Report 11 *Model procedures for the management of land contamination*, and it may be of considerable help when developing a site affected by contamination since it describes the stages of risk assessment that should be followed in order to identify risks and make judgements about the consequences of contamination for the affected site. These stages are summarized in Figure 4.2 and described in more detail below.

4.2.3 Hazard identification and assessment

One of the needs of the preliminary site assessment is to provide information on any possible contamination of the site and surrounding area due to its past

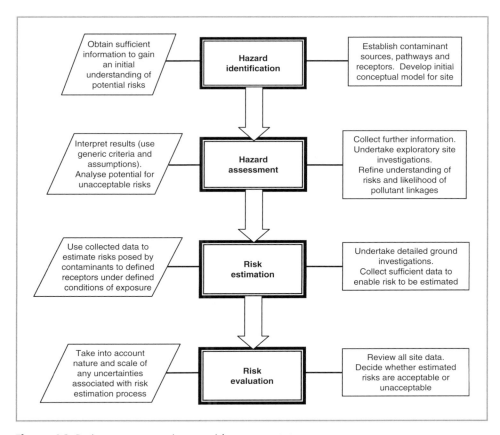

Figure 4.2 Resistance to contaminants – risk assessment stages

and present uses (see Table 2 from the AD C – reproduced here in Table 4.1 – for typical uses that might give rise to site contamination). The site walk-over survey may reveal signs of possible contaminants. Table 3 from AD C (which is reproduced in Table 4.2) gives some examples of the signs that may indicate particular contaminants; however, this is not a comprehensive list and should be used with caution since it is merely indicative.

Information provided by the desk study and site walk-over survey will be of primary importance in the design of the exploratory and detailed ground investigation.

Clearly, the site assessment and risk evaluation should be concentrated on the area of the site subject to building operations. Therefore, the following parts of a site should be remediated to the requirements of the Building Regulations:

- those parts of the land associated with the building itself
- gardens

Table 4.2 Approved Document C, Table 3 – Examples of possible contaminants

Signs of possible contaminants	Possible contaminant
Vegetation (absence, poor or unnatural growth)	Metals Metal compounds Organic compounds Gases (landfill or natural source)
Surface materials (unusual colours and contours may indicate wastes and residues)	Metals Metal compounds Oily and tarry wastes Asbestos Other mineral fibres Organic compounds including phenols Combustible material including coal and coke dust Refuse and waste
Fumes and odours (may indicate organic chemicals)	Volatile organic and/or sulfurous compounds from landfill or petrol/solvent spillage Corrosive liquids Faecal animal and vegetable matter (biologically active)
Damage to exposed foundations of existing buildings	Sulfates
Drums and containers (empty or full)	Various

- other places on the site that are accessible to users of the building and those in and about the building.

An incremental approach to remediation of a site (perhaps including lower levels of remediation) may be acceptable where part of (or the remainder of) the land associated with the building is accessible to a lesser extent to the user, or those in and about the building, than the main parts of the buildings and their respective gardens. This could also apply to areas adjacent to such land. This incremental approach may also apply in the case of phased redevelopment of very large sites where it may be possible to limit remediation to the part of the site that is actually being developed at any particular time. In all cases reliance will be placed on the risk evaluation and remediation strategy documentation in order to demonstrate that restricted remediation is acceptable. It should be noted that this will place the onus on the applicant to show why part of a site may be excluded from particular remediation measures.

The scope of the Building Regulations is limited to matters concerning health, safety, welfare and convenience; therefore, even if the adjacent land is not subject to Building Regulations, it may still be subject to planning control legislation or to control under Part IIA of the Environmental Protection Act 1990. In fact, a substantial amount of guidance on the assessment of contaminated land has been published to support the implementation of this Act. Most of this guidance is contained in the joint Defra/Environment Agency Contaminated Land Research Reports (CLRs) referred to in section 4.1.1 above. Additional guidance on these reports is given below and an outline of the process is shown in Diagram A1 from Appendix A of Approved Document C (see Figure 4.3).

Additionally, if any substance is found which is at variance with any preliminary statements made about the nature of the site, the Planning Authority should be informed before any intrusive investigations are carried out.

4.2.4 Risk estimation and evaluation

During the risk estimation phase, detailed ground investigations will be carried out. These must provide sufficient information for:

- confirmation of a conceptual model for the site
- the risk assessment itself
- the design and specification of any remedial works.

The ground investigations are likely to involve collection and analysis, by the use of invasive and/or non-invasive techniques, of:

- soil
- soil gas
- surface and groundwater samples.

Since elevated groundwater levels could bring contaminants close to the surface (both beneath the building and in any land associated with the building) an investigation of the groundwater regime, levels and flows is essential for most sites.

Therefore, if land affected by contaminants is to be developed, the health and safety of both the public and workers should be considered. Guidance on the protection of workers and the general public can be found in:

- HSE Report HSG 66 *Protection of workers and the general public during the development of contaminated land*, 1991 (23 pages). HSG 66 addresses the

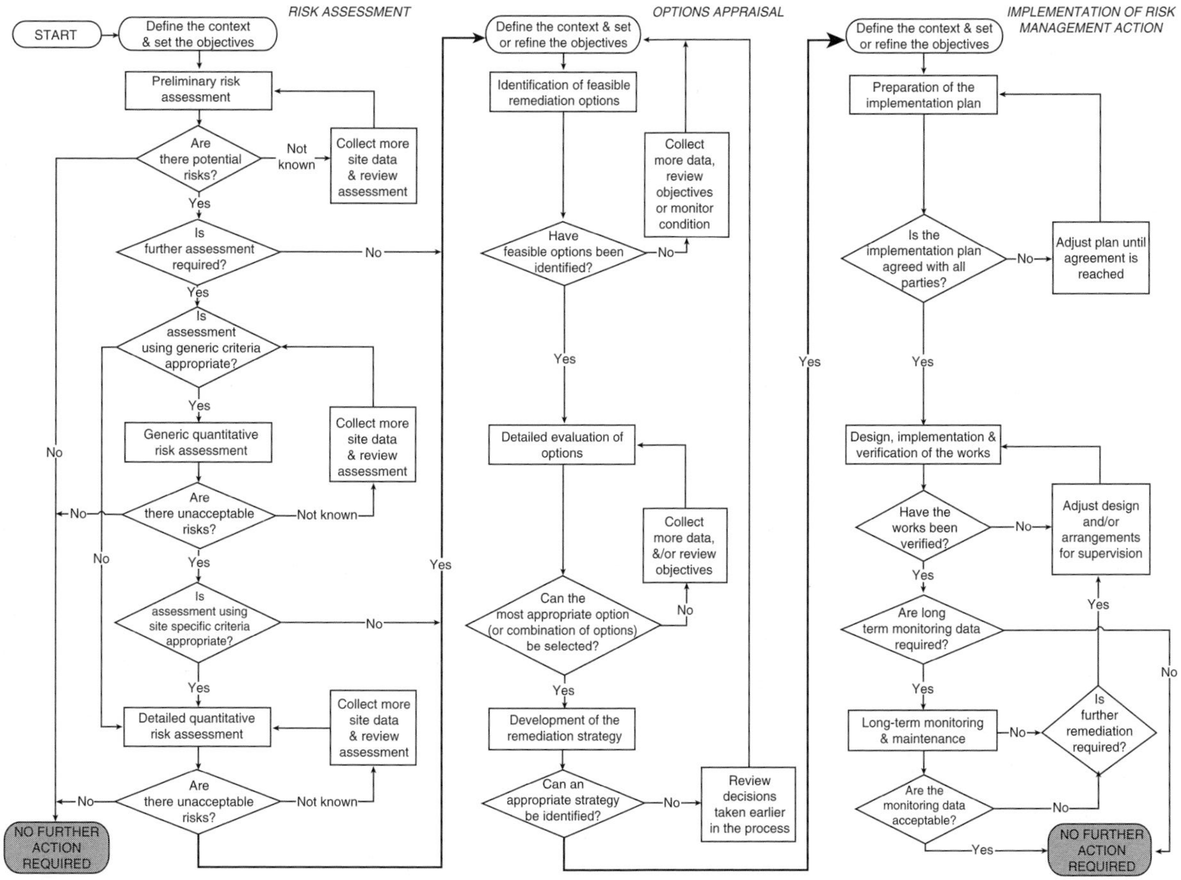

Note: The process may apply to one or more pollutant linkages each of which may follow a different route. For some linkages, it may be possible to stop at an early stage – others will progress all the way through the process. The level of complexity of each stage may also very and in some cases may be very simple.

Figure 4.3 Approved Document C – Annex A: Guidance, Diagram A1 The process of managing land affected by contaminants

considerable hazards associated with contaminated land working and aspects which developers and contractors need to consider under the Control of Substances Hazardous to Health Regulations (COSHH) and the precautions to be taken during the development of the site. It does not deal with long-term future use of contaminated land or advise on how the contamination should be treated. It will be of use to all those involved in work on contaminated land sites (e.g. surveyors, contractors, transport firms, etc.) who will need to make an assessment of the potential risks to health entailed in the work, and the precautions required to protect workers or the public. Those in control of the sites should satisfy themselves that the various contractors have carried out an assessment which is sufficient and suitable and that the specified control measures are provided and used. In most cases this assessment should be in writing.

- CIRIA Report 132 *A guide to safe working practices for contaminated land*, 1993 (239 pages). This lengthy document provides guidance on safe working practices for contaminated sites by informing those involved about:
  - their statutory responsibilities for health, safety and environmental protection
  - the potential range of contaminants and hazards and how to plan and manage the work accordingly
  - safe working methods and the use of appropriate safety clothing and equipment.

  The report covers the following main subject areas:
  - Responsibilities and obligations – including health and safety at work legislation, environmental impairment legislation, directors' responsibilities, statutory bodies and specialist consultations.
  - Hazard sources, nature and context – including COSHH assessments, hazard context, exposure routes and hazard identification information.
  - Site assessment and investigation – including preliminary assessment, investigations and surveys and guidance.
  - Planning and management of site activities – including risk assessments, design phase, construction phase, staff organization and training, monitoring and supervision, permit-to-work systems, confined spaces, reports and record-keeping, and contractual arrangements.
  - Site facilities – including site security, communication procedures, decontamination, ventilation systems and neighbourhood protection.
  - Personal protective equipment – including protective clothing and accessories, respiratory protective equipment, selection of ensembles and the use of protective clothing and respiratory protective equipment.
  - Health surveillance and first aid – including site facilities and planning, first-aid training, health surveillance of site workers and record-keeping.

The guide is intended for use by a wide readership ranging from the general engineer to contaminated land specialists. It aims to demonstrate which activities can be safely undertaken by the non-specialist and where particular expertize is required. On very heavily contaminated sites and on sites contaminated with materials subject to specific regulations (e.g. asbestos), the advice and services of experienced specialists should always be sought.

When considering health, risk estimation can be carried out using generic assessment criteria such as contaminant Soil Guideline Values (SGVs) or relevant and appropriate environmental standards. SGVs represent concentrations of contaminant which may pose unacceptable risks to health. The development of SGVs for a range of priority contaminants is described in the Defra/Environment Agency reports mentioned above. There is also a range of corresponding TOX reports which contain the toxicological data used to derive the SGVs. For example, CLR 10 describes the Contaminated Land Exposure Assessment Model (CLEA) for deriving SGVs for three different site uses:

- residential
- residential with plant uptake
- commercial/industrial.

This enables the relative importance of each of the pollutant linkages to be considered. As an example of this approach, for a residential site use the assumption is made that residents have private gardens and/or access to community open space close to the home and that some of them may use their gardens to grow vegetables. CLR 10 gives details of the conceptual model underpinning each of the standard land uses.

As an alternative to the generic approach it is also possible to undertake a more site-specific quantitative risk assessment using the principles of risk assessment or a risk assessment model. For this, specialist advice should be sought.

4.2.5 Further guidance

For guidance on the investigation of sites potentially affected by contaminants, the following documents may also be consulted:

- BS 5930: 1999 *Code of practice for site investigations* (discussed in section 3.1.1 above).
- BS 10175: 2001 *Investigation of potentially contaminated land. Code of practice* (82 pages) – This Standard provides guidance on, and recommendations for, the investigation of potentially contaminated land or land with naturally enhanced

concentrations of potentially harmful materials, to determine or manage the ensuing risks. It covers:

○ setting the objectives of an investigation
○ setting a strategy for the investigation
○ designing the different phases of the investigation
○ sampling and on-site testing
○ laboratory analysis
○ reporting

in order to obtain scientifically robust data on soil, groundwater, surface water and ground gas contamination. The relevant guidance and recommendations within the Standard need to be selected to ensure that the objectives of an investigation are achieved and that adequate data for the risk assessment are obtained. It does not provide detailed guidance for every possible investigation scenario, since this would not be feasible. Additionally, the Standard does not give recommendations on certain constraints or problems that can affect a site, such as geotechnical aspects (covered by BS 5930), or the legal aspects, including the need for licences, permits, etc. It does not include any procedures for the formal assessment of the potential risks posed by contaminated land. It is intended for use by those with some understanding of the risk-based approach to sites and site investigations, therefore it is more suitable for the specialist investigator.

• National Groundwater & Contaminated Land Centre report NC/99/38/2 *Guide to good practice for the development of conceptual models and the selection and application of mathematical models of contaminant transport processes in the subsurface.* This 120-page specialist document is aimed at hydrogeologists and environmental professionals both internal and external to the Environment Agency, who understand the concepts and processes of groundwater flow and transport in the subsurface. Its purpose is to provide guidance on a generic 'good practice' approach to contaminant fate and transport modelling from setting objectives to interpretation of results and validation. It highlights the issues that need to be considered and tackled and points to existing guidance or recognized standards and key references. It is not intended as a step-by-step recipe book for how to set up and run models. It is available from web site: www.environmentagency.gov.uk/subjects/waterres/groundwater/.

• Environment Agency R & D Technical Report P5-065 *Technical aspects of site investigation*, 2000. This two-volume document (Volume I, 101 pages and Volume II, 181 pages) provides technical guidance on the investigation of contaminated sites for use in a wide variety of contexts, including:

○ Part IIA of the Environmental Protection Act 1990
○ the planning regime
○ the Pollution Prevention and Control (PPC) regime
○ purchase or sale of land.

The document is intended to provide guidance principally to Environment Agency staff who are involved in the management of site investigation projects. However, the readership is intended to be wide ranging, including:

○ Environment Agency and local authority officers
○ those who fall under the regulatory regime and need to understand the Environment Agency's approach and requirements in relation to the investigation of contaminated sites
○ consultants and contractors engaged in site investigation projects.

The document is intended, primarily, to provide the specialist technical information required when acting in a project management capacity dealing with investigation of contaminated sites. The document should also assist in the development of a nationally consistent approach to site investigation projects in which the Environment Agency is involved by setting out what the Agency believes to be the key issues relating to good site investigation practice. The document includes an overview of good practice, technical information on the many individual investigation activities and provides a standard format for a site investigation report. The design/strategy aspects of site investigation are not dealt with in any detail within this document. The completed document provides the benefit of collective past experiences combined with 'state of the art' technical developments. It is relevant to the investigation of different contaminant types in ground and groundwater on all sites where contamination investigation is an issue.

• Environment Agency R & D Technical Report P5-066 *Secondary model procedure for the development of appropriate soil sampling strategies for land contamination* (105 pages). The document comprises two parts:

○ Part A contains explanatory material that discusses the relationship between soil sampling and risk assessment, the importance of the Conceptual Model, uncertainty and data quality, and key design parameters.
○ Part B contains procedures for developing appropriate soil-sampling strategies and for reviewing planned or completed work, and addresses only those matters relevant to the development of sampling strategies for soils and closely allied materials such as solid wastes, ground gases and vapours, the leachable fraction, soil pore waters and non-aqueous phase liquids.

This specialist document is aimed at three main user groups:

○ those who design soil sampling strategies, usually as part of a wider site characterization exercise
○ those who use factual information about the condition of a site and its setting to assess health and environmental risks
○ those who rely on the output of the first two groups and who need to be satisfied about the technical validity of the work done.

These documents recommend the adoption of a risk-based approach, so that any hazards that are present can be identified and quantified and an assessment can be made of the nature of the risk they might pose. The design and execution of field investigations are described together with suitable sample distribution strategies, testing and sampling.

4.3 Resistance to contaminants – remedial measures

4.3.1 Introduction

If the risk assessment stage results in the identification of unacceptable risks to the particular receptor (e.g. buildings, building services, building materials and people), then appropriate remedial measures will be needed in order to manage these risks. This will mean defining the risk management objectives in terms of the need to break the pollutant linkages. Other objectives that will also need to be considered include such items as:

- timescale and cost
- remedial works
- planning constraints
- sustainability.

The remedial measures that are adopted will depend on the contaminant that has been identified. In general, three generic types of remedial measures can be considered:

- treatment
- containment
- removal.

It should be noted that it may be necessary to obtain a waste management licence from the Environment Agency where the containment or treatment of waste is anticipated. It is also important to consider the affects of building work on sites affected by contaminants. For example an existing control measure which included a cover system (i.e. containment) could be breached if excavations were carried out for foundations or underground services when an extension was added.

4.3.2 Treatment

Contaminants can be dealt with by a wide range of treatment processes using biological, chemical and physical techniques. These can be carried out either on or off the site and are designed to decrease one or more of the contaminant's features such as mass, concentration, mobility, flux or toxicity. Since the choice of the most appropriate technique is highly site-specific, specialist advice should be sought.

4.3.3 Containment

In general, the term 'containment' means encapsulation of material containing contaminants. However, in the context of building development it is usually taken to mean cover systems, sometimes incorporating vertical barriers in the ground to control lateral migration of contaminants.

In a cover system, layers of materials are placed over the site in order to:

- interrupt the pollutant linkage between the contaminants and the receptors
- sustain vegetation
- improve geotechnical properties
- reduce exposure to an acceptable level.

Some parts of the structure of the building (foundations, substructure, ground floor, etc.) may assist other containment measures in providing effective protection of health from contaminants. However, the extent to which this is feasible will depend on the circumstances and form of construction.

The following issues need to be addressed when using imported fill and soil for cover systems:

- It should be assessed at source to ensure that it is not contaminated above specified concentrations.
- It should meet required standards for vegetation (see BS 3882: 1994 *Specification for topsoil*. This 34-page Standard specifies requirements for topsoils. It establishes three grades of material and gives recommendations for the use and handling of topsoil. It is not intended (or appropriate) for the grading, classification or standardization of *in situ* topsoil or subsoil. Methods of sampling are given in Annex A in the Standard. Methods for determination and calculation of various compounds present in topsoil are given in Annex B to Annex M and Annex P. The Standard is suitable for specialists in the analysis of soils).
- Where intermixing of the soil cover with the contaminants in the ground can take place it will be necessary in the design and dimensioning of the cover system to consider its long-term performance. This may include the need for maintenance and monitoring.

- Gradual intermixing of the soil and contaminants due to natural effects and activities (e.g. by burrowing animals, gardening, etc.) should be taken into account.
- Excavations by householders for garden features, walls, ponds, etc., can penetrate the cover layer leading to possible exposure to contaminants.

Further guidance on the design, construction and performance of cover layers can be found in the Construction Industry Research and Information Association (CIRIA) Special Publication SP124 *Barriers, liners and cover systems for containment and control of land contamination*, 1996 (280 pages). The publication explains that containment and control can be achieved either by constructing suitable in-ground barriers or by using hydraulic measures. Barrier systems often have to cope with both liquid and gaseous contamination. SP124 covers design criteria, theory and practice for the full range of physical barriers. Guidance is given relating to in-ground barriers, multi-layer cover systems and liners, the theoretical basis for these systems and appropriate *in situ* and laboratory testing and analysis, for design and performance monitoring. Experience with physical containment is reviewed and presented in terms of the efficacy of these techniques, the need for close quality control over materials and installation and how they may be improved.

The structure of SP124 attempts to satisfy both the need for detailed information and to describe the background and concepts necessary for the effective and safe design and construction of in-ground physical barriers to contain and isolate contamination as part of a remedial strategy for achieving beneficial reuse of land. The need for, and use of, monitoring and the wider implication of the duty of care (as defined for the UK by the Environmental Protection Act 1990) is discussed.

In general, the report reflects 'good practice' not only in the context of current policy and regulations but also as they are expected to develop in the foreseeable future and is applicable to most contaminated land situations including those which do not conform with the classic 'redevelopment' scenario, such as highways and active industrial sites.

It is designed to meet the needs of a wide range of potential users, including project and development managers, consultants and contractors acting on behalf of public and private development agencies, other clients of the construction industry, central and local government, and other regulatory authorities.

4.3.4 Removal

Removal means the excavation and safe disposal to a licensed landfill site of the contaminants and contaminated material. This can be achieved by targeting the excavation on contaminant hot spots, or by removing sufficient depth of contaminated material so that a cover system can be accommodated within the planned site levels. Removal may not always be viable since it will depend on the depth and extent of the contaminants on the site and the availability of suitably licensed landfills.

Where removal is incorporated with a subsequent cover system any imported fill should be assessed at source to ensure that there are no materials that will pose unacceptable risks to potential receptors.

Further detailed guidance on treatment, containment and removal is given in the Environment Agency/NHBC R & D Publication 66 referred to in section 4.2.5 and in the following CIRIA publications:

- CIRIA Special Publication SP102 *Decommissioning, decontamination and demolition*, 1995
- CIRIA Special Publication SP104 *Classification and selection of remedial methods*, 1995
- CIRIA Special Publication SP105 *Excavation and disposal*, 1995
- CIRIA Special Publication SP106 *Containment and hydraulic measures*, 1996
- CIRIA Special Publication SP107 *Ex-situ remedial methods for soils, sludges and sediments*, 1995
- CIRIA Special Publication SP109 *In-situ methods of remediation*, 1995.

In fact these publications (which are specifically referred to in Approved Document C) form part of a 12-volume set published by CIRIA under the general heading *Remedial treatment for contaminated land*.

All the volumes in this series of publications are listed in Table 4.3.

The report (i.e. all 12 volumes) is intended for:

- owners of contaminated sites proposing to take remedial action as a prelude to sale or disposal of land and property, as a precursor to redevelopment, as part of a corporate environmental management programme, or because action is needed to avert a public health and/or environmental threat
- non-specialist managers who, faced with redeveloping or remediating contaminated sites, need information and guidance for procurement and project management purposes
- non-specialist civil engineering, architectural or construction advisers providing design, supervision and inspection services in collaboration with specialist advisers
- contracting organizations providing groundworks, drilling, surveying, landscape, laboratory analysis, waste management services, etc. to remediation projects

Table 4.3 Twelve-volume set of CIRIA *Remedial treatment for contaminated land* publications

Volume	Title	Outline content	Notes
I (SP101)	Introduction and guide	Aims, scope, contents list and brief summaries	
II (SP102) (124 pages)	Decommissioning, decontamination and demolition	Issues to be addressed and guidance on procedures	Referred to in AD C (see summary at the end of section 4.3.4)
III (SP103)	Site investigation and assessment	Issues to be addressed and guidance on procedures	
IV (SP104) (176 pages)	Classification and selection of remedial methods	Classification and selection of appropriate methods and strategies	Referred to in AD C (see summary at the end of section 4.3.4)
V (SP105) (55 pages)	Excavation and disposal	Description and evaluation of methods and guidance on procedures	Referred to in AD C (see summary at the end of section 4.3.4)
VI (SP106) (139 pages)	Containment and hydraulic measures	Description and evaluation of methods and guidance on procedures	Referred to in AD C (see summary at the end of section 4.3.4)
VII (SP107) (187 Pages)	*Ex situ* remedial methods for soils, sludges and sediments	Description and evaluation of methods and guidance on procedures	Referred to in AD C (see summary at the end of section 4.3.4)
VIII (SP108)	*Ex situ* remedial methods for contaminated groundwater and other liquids	Description and evaluation of methods and guidance on procedures	
IX (SP109) (190 pages)	*In situ* methods of remediation	Description and evaluation of methods and guidance on procedures	Referred to in AD C (see summary *below*)
X (SP110)	Special situations	Information and guidance relating to situations outside the redevelopment context	
XI (SP111)	Planning and management	Issues to be addressed and guidance on procedures	
XII (SP112)	Policy and legislation	Information on policy, administration and legal frameworks in UK and overseas	

- regulatory bodies having responsibility for public and occupational health and safety, and protection of the environment, at all stages of managing a contaminated site.

The report assumes that readers will have a basic understanding of the nature of the problems of land contamination.

It is interesting that only selected volumes are referred to in Approved Document C since CIRIA makes it clear in its Foreword to each volume that the entire report is intended to be used '*as a single source of information and guidance on the assessment and remediation of contaminated sites*'. It goes on to say that '*Although each Volume is self-contained to the extent that it covers the principal issues and procedures relevant to the subject area, reference to other Volumes may be necessary for more detailed information and discussion on specific aspects. Extensive cross referencing between the various Sections and Volumes is provided to help users locate this more detailed information where necessary*'.

Therefore, it would be prudent for users falling within the above categories to obtain the full report. The contents of the volumes referred to in AD C are outlined as follows:

- SP 102 (Volume II) deals with design and implementation of post-closure operations for contaminated sites. Planning, monitoring, health and safety, environmental protection and other post-treatment management tasks are covered. Emphasis is placed on the importance of site surveys before and after decommissioning, decontamination and site clearance, on problems related to public health and the environment, and on the additional costs that may be incurred through poor planning and execution of site clearance operations.
- SP104 (Volume IV) provides a classification of the techniques available for the treatment of contaminated sites. Their typical characteristics and requirements are described. A step-by-step approach to the selection of appropriate techniques is given, this includes setting objectives, identifying constraints, evaluating site information and testing of various treatment options against the selection criteria.
- SP105 (Volume V) deals with the excavation of contaminated material prior to disposal (on- or off-site) or as a precursor to other forms of treatment. Issues considered include the applicability, limitations, effectiveness and practical requirements of excavation, and the technical, administrative and legal implications of on and off-site disposal.
- SP106 (Volume VI) provides information and guidance on engineering-based remedial methods, specifically physical containment and hydraulic control measures. These include cover layers, vertical and horizontal barriers, liner systems, control or isolation of contaminated groundwater, management of the hydraulic

regime, maintenance of favourable hydraulic gradients across physical barriers, and removal of contaminated groundwater from a site. Technical and operational parameters are considered and supporting examples and references are provided.

- SP107 (Volume VII) deals with physical, chemical and biological methods of removing or rendering harmless the contaminants in solid materials after excavation from the ground. Guidance is given on the requirements for equipment, process controls and ongoing monitoring, and on material handling procedures. The effectiveness, limitations and practical implementation of the various techniques are discussed.
- SP109 (Volume IX) describes the techniques available for removing, destroying or rendering harmless contaminants while they are in place in the ground. The applications to which they are suited, and their operational characteristics and requirements are considered and compared with *ex situ* remedial treatments.

4.4 Resistance to contaminants – risks to buildings, building materials and services

Receptors of contaminants include not only people but also buildings, building materials and services on sites. The hazards to these receptors might include:

- **Aggressive substances** – which may affect the long-term durability of construction materials such as concrete, metals and plastics (e.g. organic and inorganic acids, alkalis, organic solvents and inorganic chemicals such as sulphates and chlorides).
- **Combustible fill** – which may lead to subterranean fires, if ignited, and consequent damage to the structural stability of buildings, and the integrity or performance of services (e.g. domestic waste, colliery spoil, coal, plastics, petrol-soaked ground, etc.).
- **Expansive slags** – which may expand some time after deposition (usually when water is introduced onto the site) causing damage to buildings and services (e.g. blast furnace and steel making slag).
- **Contaminant-affected floodwater** – floodwater may be contaminated by substances in the ground, waste matter or sewage. Building elements that are close to or in the ground, such as walls or ground floors may be affected by this contaminated water. The following documents contain guidance on resistant construction:
  - ○ *Preparing for floods: interim guidance for improving the flood resistance of domestic and small business properties*, DTLR, 2002 (100 pages). The aim of the guide is to show how the flood resistance of properties may be improved.

Sections 1 and 2 provide a general introduction and some background information on the causes and impact of flooding. Section 3, which is aimed principally at existing homeowners and small business owners provides information on assessing the risks of flooding, and guidance on selecting appropriate measures to improve the flood resistance of properties. Information is also provided on measures to prevent or reduce the volume of floodwater entering the building, including the use of temporary flood barriers and other permanent measures to improve the flood resistance of the building structure. Section 3 also provides outline advice on the special considerations that apply to properties of special architectural or historic interest, and gives references to more detailed guidance provided by English Heritage. Section 4 is for use by developers, local authorities, building control bodies and others involved with new development in high flood risk areas. It provides guidance on the forms of construction that are most appropriate for developments at risk of flooding. Section 5 provides more technical information on the permanent measures that can be taken to improve the flood resistance of both existing and new buildings with subsections discussing walls, floors and building services and fittings (such as electrical wiring and fitted cupboards). This section is aimed principally at builders but may also be of interest to property owners. Section 6 provides a summary of the key steps to reducing the consequences of flooding as outlined within the guide. Section 7 includes a list of related publications for further reading, and Section 8 gives guidance from the Environment Agency's Floodline service. Case studies are included throughout the guide to give real-life examples of what steps homeowners and small business owners have taken to protect their properties from flooding. The guide is intended for use by property owners, developers, local planning authorities and others involved in construction of new buildings, and renovation of existing buildings, where their buildings are at risk of flooding.

○ *BRE for Scottish Office Design guidance on flood damage for dwellings*, TSO, 1996 (31 pages). This guide covers issues affecting design (i.e. risk of flooding, site considerations, entry of water into dwellings, floodwater containing salt, silt or sewerage, wetting and drying, materials and constructions); the effects of water on materials (masonry and concrete, timber, wall finishes, metals, insulation); guidance on construction and details (including ground supported floors, suspended timber floors, suspended concrete floors, solid and cavity masonry walls, framed walls, secondary elements such as partitions, walls and doors, finishes, building services and fittings). Section 5 of the guide contains sources of further information. This is a simple, eminently practical guide, which will be of great use to anyone involved in the design, construction, renovation, repair and alteration of houses in flood risk areas.

Although the main receptors with these hazards are the building, the building materials and the building services, ultimately the health of the occupants may be put at risk. In particular, potable water pipes made of polyethylene may be permeated by hydrocarbons.

Additionally, the Environment Agency document *Assessment and management of risks to buildings, building materials and services from land contamination*, 2001 (98 pages), contains further guidance. This document considers the risks to buildings, building materials and services that may be associated with land contamination in terms of four principal hazards:

- the presence of aggressive substances (e.g. inorganic and organic acids, alkalis, organic solvents, and inorganic salts such as sulphates and chlorides, which may affect the long-term performance and durability of construction materials used in contaminated ground)
- the presence of combustible materials which, if ignited may lead to subterranean fires and consequent damage to the structural stability of buildings, and the integrity/performance of site services
- the presence of expansive slags, which has implications for structural stability and serviceability
- the presence of unstable fills, which has implications for structural stability and serviceability.

The document advises how to:

- identify the potential hazards mentioned above
- carry out the site investigation and associated analysis and testing necessary to properly assess the risks
- assess the risks posed by the hazards including use of relevant generic assessment criteria, where available

and is aimed at all those involved in the management of land contamination including regulators, specialist consultants and contractors, construction clients, developers, main contractors, sub-contractors, consulting engineers and other construction professionals, and producers of construction materials. It will also enable a wide audience to appreciate the issues involved.

4.5 Radon gas contamination

4.5.1 Introduction

Radon is a naturally occurring, colourless and odourless gas which is radioactive. It is formed in small quantities by the radioactive decay of uranium and radium, and thus travels through cracks and fissures in the subsoil until it reaches the atmosphere or enters spaces under or in buildings.

It is recognized that radon gas occurs in all buildings; however the concentration may vary from below 20 Bq/m^3 (the national average for houses in the UK) to more than 100 times this value. The National Radiological Protection Board (NRPB) has recommended an action level of 200 Bq/m^3 for houses. The lifetime risk of contracting lung or other related cancers at the action level is about 3%. Geographical distribution of houses at or above the action level is very uneven, with about two-thirds of the total being in Devon and Cornwall.

The Office of the Deputy Prime Minister is reviewing the areas where preventative measures should be taken as information becomes available from the NRPB and the British Geological Survey (BGS). This information has been placed in the Building Research Establishment (BRE) guidance document *Radon: guidance on protective measures for new dwellings* (BRE Report BR 211: third edition 1999) (54 pages), obtainable from Building Research Establishment, Garston, Watford, WD2 7JR) which will be updated as necessary.

A common basis for radiation protection legislation in all member states of the European Union has been established in a European Council Directive. This Directive has been put into effect in the UK by virtue of the *Ionising Radiations Regulations 1999* (SI 1999/3232). These regulations set a national reference level for a range of radioactive substances including radon gas and they require employers and self-employed persons responsible for a workplace to measure radiation (including radon) levels on being directed to do so.

Reference may also be made to BRE Report BR 293 *Radon in the workplace*, 1995, which provides guidance for existing non-domestic buildings. This 51-page report deals with:

- radon and its health effects
- legislation including the employer's responsibility in the workplace (see SI 1999/3232 above). (Care should be taken when using this section since the report predates the 1999 Regulations and discusses the earlier 1985 Ionising Radiations Regulations.)
- measurement of radon in the workplace
- identifying the extent of the problem where it has been recognized that a building has an elevated indoor radon level
- background information on the building obtained by speaking to the building owner, manager or user
- construction survey of the property
- choosing a solution – which may be:
  - generic (sealing, positive pressurization, sumps, underfloor ventilation, ventilation (other than positive pressurization), or

- ○ solutions for complicated situations (e.g. where there are floors of mixed construction, basements or stepped construction)
- who should carry out the survey of the building and installation of remedial work
- retesting for radon
- additional protection for landfill gas
- new buildings, extensions and major alterations and conversion works.

It also contains a useful appendix with 12 case studies.

The report is aimed principally at employers and those who control buildings used for work purposes, or their representatives. The guidance should also be of interest and assistance to those, such as surveyors and builders, concerned with specifying and carrying out the necessary remedial measures.

The guidance in BR 211 has been developed to show radon protective measures in dwellings. No guidance exists at present for radon protection in new workplaces (BR 293 deals with existing workplaces). However, for domestic-sized workplaces with heating and ventilation regimes similar to those in dwellings (such as small office buildings and primary schools) some of the techniques described in BR 211 for installing radon-resistant membranes, may be suitable. The guidance in BR 211 can also be used as the basis for radon protection of other building types provided that caution is exercised.

Although no guidance is currently available in the 1999 edition of BR 211 for suspended timber ground floors in new dwellings, it is understood that the ODPM is sponsoring research into how this form of construction could provide adequate protection against radon.

BR 211 identifies those areas where either basic or full radon protection is needed by reference to a series of maps derived from:

- statistical analysis of radon measurements of existing houses carried out by the NRPB (Annex A of BR 211)
- an assessment of geological radon potential prepared by the British Geological Survey (Annex B of BR 211).

Use of the maps in accordance with directions given in BR 211 will determine whether basic, full or no protection is needed.

Areas most at risk include parts of:

- Devon
- Cornwall
- Somerset
- Gloucestershire

- Oxfordshire
- Northamptonshire
- Leicestershire and Rutland
- Lincolnshire
- Staffordshire
- Derbyshire
- West Yorkshire
- Northumberland and parts of southern Cumbria
- most of Wales.

As more information becomes available from the NRPB it is likely that further areas will be covered by the need for radon precautions. Current information on the areas delineated by ODPM for the purposes of Building Regulations can be obtained from local authority building control officers or from approved inspectors. When it is necessary to make changes to areas delineated as requiring radon protection these will be notified to building control bodies and will be posted on the ODPM web site. The results will be published in due course.

Interim guidance on radon measures in domestic buildings (including conservatories and extensions) may be found in Good Building Guide 25 *Buildings and radon*, 1996 (12 pages). The guide is divided into four sections and supplements existing guidance by drawing together different areas of BRE radon-related research. The first part of the guide describes the benefits of passive sump systems, i.e. systems that are not fan assisted (sump systems are usually very effective at reducing indoor radon levels). The second part shows how a single fan-assisted system can be used to treat several adjoining houses. Compared to installing several separate systems, a communal system of this type is quicker and cheaper to install and causes less disruption. Other sections describe how to safeguard against the spillage of combustion products when using a radon remedial system, and how to protect new extensions and conservatories against the entry of radon. Papers in the Good Building Guide series produced by the BRE are aimed at non-specialists. The material is presented in an easily readable form with minimal technical jargon and plenty of simple illustrations. Anyone requiring more detailed information on radon protection should consult the other reference sources mentioned in this section.

4.5.2 Basic protection against radon

Basic protection may be provided by an airtight, and therefore radon-proof, barrier across the whole of the building including the floor and walls. This could consist of:

- polyethylene (polythene) sheet membrane of at least 300 micrometre (1200 gauge) thickness
- flexible sheet roofing materials
- prefabricated welded barriers
- liquid coatings
- self-adhesive bituminous-coated sheet products
- asphalt tanking.

It is important to have adequately sealed joints and the membrane must not be damaged during construction. Where possible, penetration of the membrane by service entries should be avoided. With careful design it may be possible for the barrier to serve the dual purpose of damp-proofing and radon protection, although the damp-proof course to a cavity wall should be in the form of a cavity tray to prevent radon entering the building through the cavity.

Some typical details are shown in Figure 4.4.

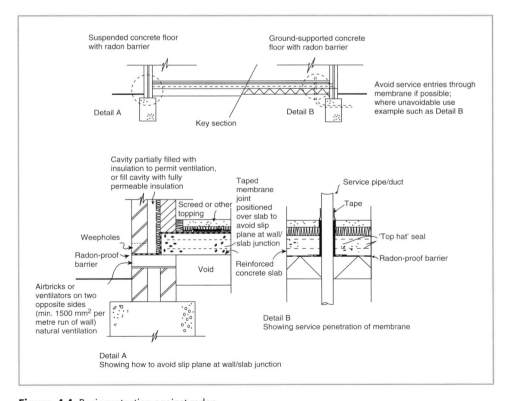

Figure 4.4 Basic protection against radon

4.5.3 Full protection against radon

In practical terms, a totally radon-proof barrier may be difficult to achieve. Therefore, in high-risk areas it is necessary to provide additional secondary protection. This might consist of:

- natural ventilation of an underfloor space by airbricks or ventilators on at least two sides
- the addition of an electrically operated fan in place of one of the airbricks to provide enhanced subfloor ventilation
- a subfloor depressurization system comprising a sump located beneath the floor slab, joined by pipework to a fan. It may only be necessary to provide the sump and underfloor pipework during construction thus giving the owner the option of connecting a fan at a later stage if necessary.

Examples of these methods are shown in Figures 4.4 and 4.5.

It should be noted that the above brief notes on BR 211 are intended to give an idea of the content of that document. Designers of buildings in the delimited areas should consult the full report.

4.6 Contamination of landfill gas

4.6.1 Introduction

Landfill gas is typically made up of 60% methane and 40% carbon dioxide, although small quantities of other gases such as hydrogen, hydrogen sulphide and a wide range of trace organic vapours (called volatile organic compounds or VOCs in Approved Document C) may also be present. The gas is produced by the breakdown of organic material by micro-organisms under oxygen-free (anaerobic) conditions on biodegradable waste materials in landfill sites. Gases similar to landfill gas can also arise naturally from coal strata, river silt, sewage and peat. Additionally, atmospheres which are deficient in methane and oxygen (usually referred to as stythe or black-damp by miners) and which are rich in carbon dioxide and nitrogen can be produced naturally in coal-mining areas. Volatile organic compounds can also arise as a result of spillages of petrol, oil and solvents.

4.6.2 Properties of landfill gases

The largest component of landfill gas, methane, is a flammable, asphyxiating gas with a flammable range between 5% and 15% by volume in air. If such

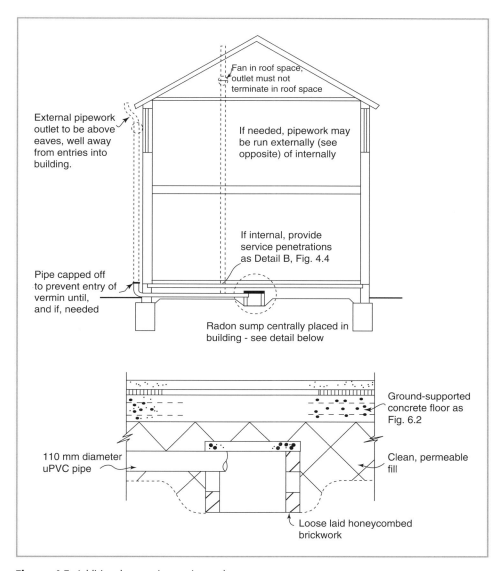

Fan in roof space, outlet must not terminate in roof space

External pipework outlet to be above eaves, well away from entries into building.

If needed, pipework may be run externally (see opposite) of internally

If internal, provide service penetrations as Detail B, Fig. 4.4

Pipe capped off to prevent entry of vermin until, and if, needed

Radon sump centrally placed in building - see detail below

Ground-supported concrete floor as Fig. 6.2

110 mm diameter uPVC pipe

Clean, permeable fill

Loose laid honeycombed brickwork

Figure 4.5 Additional protection against radon

a concentration occurs within a building and the gas is ignited, it will explode. Methane is lighter than air.

The other major component, carbon dioxide, is a non-flammable, toxic gas which has a long-term exposure limit of 0.5% by volume and a short-term exposure limit of 1.5%. It is heavier than air. Volatile organic compounds are both inflammable and toxic, and can have strong unpleasant odours. A build-up to hazardous levels of any of these gases within buildings will result in harm to health and will compromise safety.

4.6.3 *Movement of landfill gases*

The proportions of the two main landfill gases and the amount of air mixed with them will largely determine the properties of the landfill gas since they remain mixed and do not separate, although the mixture can remain separate from surrounding air. These landfill gases will migrate from a landfill site as a result of diffusion through the ground and this migration may be increased by rainfall or freezing temperatures as these conditions tend to seal the ground surface. The gases will also follow cracks, cavities, pipelines and tunnels, etc., as these form ideal pathways. Landfill gas emissions can be increased by rapid falls in atmospheric pressure and by a rising water table. Thus, landfill gas may enter buildings and may collect in underfloor voids, drains and soakaways.

4.6.4 *Building near landfill sites and on gas contaminated land – risk assessment*

The risk assessment stages referred to in section 4.2.2 above should be followed for methane and other landfill gases. Further investigations for hazardous soil gases may also be required if the ground to be covered by a building and/or any land associated with the building is:

(a) on, or within 250 m of a landfill site or within the likely sphere of influence of a landfill. In these cases the policy of the Environment Agency regarding building on or near landfill sites should be followed
(b) on a site where biodegradable substances (including made ground and fill) have been deposited on a large scale
(c) on a site where the previous use has meant that spillages of petrol, oil and solvents could have taken place (such as vehicle scrap yards)
(d) in an area where naturally occurring gases (e.g. methane, carbon dioxide, hydrogen sulphide and VOCs) may be present (such as old mining areas and spoil heaps).

In these specific instances Approved Document C recommends use of the following guidance documents covering hazardous soil gases in the contexts referred to in (a) to (d) above:

• Guidance on the generation and movement of landfill gas as well as techniques for its investigation are given in HMIP Waste Management Paper No 27 *The control of landfill gas*, TSO, 2nd edition 1991 (82 pages). The paper discusses the main factors responsible for the formation of landfill gas and goes on to describe:

○ the properties of the gas mixture
○ factors affecting its migration
○ methods of assessing, monitoring and controlling the gas

in order to give information on the options available for its management. The paper does not consider the management and control of leachate, even though it is recognized to be closely associated with gas. The paper is written in non-technical terms and is a useful way of gaining an understanding of the behaviour, monitoring and control of landfill gas. Appendix B on *Main relevant legislation* should be viewed with caution due to the publication date of the paper (1991) since much of it has now been superseded. The paper would be of use to anyone concerned with the development of land on or near landfill sites. Complementary guidance is given in a publication by the Chartered Institution of Wastes Management (CIWM) entitled *Monitoring of landfill gas*, 2nd edition 1998. This guidance document provides detailed information on all aspects of gas generation and monitoring under a variety of conditions. Case studies are presented to illustrate the problems associated with monitoring landfill gas. The document also provides useful information on the appropriateness of specific monitoring techniques and methods for distinguishing landfill gas from other sources of gas.

- Guidance on the geographical extent, associated hazards and methods of site investigation of methane, carbon dioxide and oil seeps from natural sources and mining areas can be found in BGS Technical Report WP/95/1 *Methane, carbon dioxide and oil seeps from natural sources and mining areas: characteristics, extent and relevance to planning and development in Great Britain*, 1995. This should be read in conjunction with a report sponsored by the former Department of the Environment entitled *Methane and other gases from disused coal mines: the planning response*, TSO, 1996 (72 pages). Under certain circumstances, hazardous mine gases can seep to the surface from abandoned underground coal workings. Such emissions have affected reclamation and construction sites, residential, commercial and industrial properties. Fatalities and injuries have occurred but they are rare. The study which resulted in the publication of this report was aimed at identifying a suitable planning response to reduce mine gas emission risks in respect of new development, without placing unnecessary constraints on land use. The study also has relevance to the detection, investigation and treatment of mine gas emissions affecting existing development. The report will provide advice suitable for use by planners, developers, land and property owners, insurers and others, in current and past coal-mining areas.

- Additionally, the following three documents published by CIRIA (which are part of CIRIA's continuing programme on *Methane and associated hazards to construction*) contain relevant information on methane and other gases including means of generation and movement within the ground, detection and monitoring methods, and investigation strategies (in all five reports in this series are referred to below).

- o CIRIA Report 130 *Methane: its occurrence and hazards in construction*, 1993 (139 pages). The report starts with a summary of the physical and chemical properties of methane and other landfill gases and explains their hazardous characteristics. An explanation is given of how and in what situations methane is formed, how it moves or can be moved in the ground, and how the source of the methane can be identified. Case histories and scenarios of typical situations are used to show how and where methane can affect construction projects. The report and its appendices provide information which will enable construction professionals to recognize potential methane problems and to initiate the process of finding solutions for them. Engineering solutions, however, are not put forward as these are the subject of subsequent CIRIA projects. The report is aimed at providing guidance for construction professionals who may have to take methane (and other gases often present with it) into account during the construction process. Although there are some sections of general interest, much of the report requires advanced knowledge of chemistry and mathematics.

- o CIRIA Report 131 *The measurement of methane and other gases from the ground*, 1993 (99 pages). This report provides guidance on the detection, measurement and monitoring of gases in the ground. Although the report is centred on methane, other hazardous gases such as carbon dioxide, hydrogen sulphide, carbon monoxide and hydrogen may be present with methane, or occur separately, and the report provides guidance for these gases. Within the context of safe working, comprehensive guidance is given on detecting gas, identifying the source, measuring and sampling different gases, and on the interpretation of the results. Factors which affect the investigation of gas, such as meteorological conditions, gas in groundwater, etc. are discussed. The use of monitoring systems on a site and within buildings is described. An appendix lists standards and codes relevant to safety in site investigations for gas. This should be used with caution since many of the references are out of date due to the age of the publication (1993). The report will be of most use to specialist companies directly involved in the detection, measurement and monitoring of gases from the ground, and for those people responsible for commissioning their services.

- o CIRIA Report 150 *Methane investigation strategies*, 1995 (84 pages). This report gives guidance for good practice in the design and execution of site investigations for methane and associated gases in the ground. The report was prepared following a comprehensive review of current guidance documents and detailed consultation with practitioners, developers, local authorities, funders and insurers. The guidance given is intended to be indicative rather than prescriptive, and the importance of site specificity is emphasized. The report gives an account of the literature review and consultations, and describes good practice for site investigations with the aid of tables, a flowchart and examples. Emphasis is given to the procedures and strategies which should be adopted rather than specific techniques, although a brief account of these is also given. The report is of a specialist nature and will be of most use to those people responsible for commissioning methane investigations.

When a site investigation is carried out for methane and other hazardous gases, consideration should be given to the following matters:

- In order fully to characterize gas emissions, measurements should be taken over a sufficiently long period of time (including periods when gas emissions are likely to be higher, e.g. during periods of falling atmospheric pressure).
- It is also important to establish:
  - the concentration of methane and other gases in the ground
  - the quantity of gas generating materials and their rate of gas generation
  - gas movement in the ground
  - gas emissions from the ground surface.

Measurements taken of the surface emission rates and borehole flow rates will give an indication of the gas regime in the ground. For further guidance on this reference should be made to:

- CIRIA Report 151 *Interpreting measurements of gas in the ground*, 1995 (103 pages). This report reviews the limitations of current gas measurement techniques and recommends ways to standardize and improve not only the techniques of measurement, but also the ways to develop sound interpretations. Its purpose is to assist those planning, undertaking and interpreting gas investigations to take sensible measurements and to make sense of the measurements taken. The report shows how the systems of measurement affect the values measured, how external conditions alter the gas regime and how, by recognizing what has and has not been measured, the results of the measurements can be interpreted in the site context. Since the report contains highly technical information, it will be of most use to engineers allowing them to test the validity of gas measurements and their meaning.
- CIRIA Report 152 *Risk assessment for methane and other gases from the ground*, 1995 (70 pages). This report proposes a rational methodology for gas hazard evaluation and risk assessment appropriate for a wide range of construction

situations and ground gases (but particularly focusing on methane and carbon dioxide). The report outlines the principles of risk assessment and applies them to the context of ground gas entering a building thus relating the risks presented by these gases to other tolerable risks and specifically to those of natural gas supply mains. Whilst this document is mainly for the use of construction professionals, (it sets out good practice in gas hazard evaluation and risk assessment and shows how these can be applied to particular situations) much of it is non-technical and will be of use to all those concerned with risk assessments of sites where the presence of ground gases is suspected, since it draws together the guidance given in the earlier CIRIA reports in this series on factors contributing to the risk. It also acts as a guidance document to assessors of risk of gas ingress in development situations.

The gas regime on the site can also be altered by construction activities undertaken as part of building development. For example, surface gas emissions can be increased by the site strip, by piling, and by excavations for foundations and below ground services; and dry biodegradable waste can be pushed into moist, gas-active zones by dynamic compaction.

Reference was made in section 4.2.4 above to the use of contaminant Soil Guideline Values (SGVs) when estimating the risk to health and safety of the public on a site. It should be noted that there are no Soil Guideline Values for methane and other gases.

Therefore, in the context of traditional housing the assessment of gas risks needs to take into account two possible contaminant pathways for human receptors:

(a) direct entry of gas into the dwelling through the substructure (where it will ultimately build up to hazardous levels)
(b) later exposure of the householder in garden areas by:
  (i) the construction of outbuildings (e.g. garden sheds and greenhouses) and extensions to the dwelling
  (ii) the carrying out of excavations for garden features (e.g. ponds).
  Guidance on the carrying out of risk assessments for methane and other ground gases can be found in CIRIA Report 152 referred to above and in Contaminated Land Research Report CLR 11 (see section 4.1.1 above).

The following documents describe a range of ground gas regimes (defined in terms of soil gas concentrations of methane and carbon dioxide as well as borehole flow rate measurements) which can also be helpful in assessing gas risks:

- CIRIA Report 149 *Protecting development from methane*, 1995 (192 pages). This report examines the need for buildings to be protected from hazards arising from methane and other gases. It sets out the principles for providing adequate protection in different situations of gas regime and by building type. The report describes the specific components of a protective system which include:

- in-ground barriers (vertical, horizontal and of different materials)
- in-ground venting and active abstraction of gas
- venting for buildings
- the installation of gas membranes
- the operation and management of gas monitoring and alarm systems.

The range of gas-control systems used for different gas regimes in the ground is shown and the particular needs and constraints of providing protection to existing buildings, services and substructures are described. Guidance is given on the long-term management of gas-control systems. This report provides a practical guide to current accepted good practice on the selection, design and performance of methane protection measures for new and existing building development including associated works. Since the comments made in the report are based on information gathered from various sources (i.e. from consultations with experts, practitioners, clients and owners of development, public utility engineers and those with a statutory or regulatory responsibility for development) it will be of great use to all these readers.

- DETR/Arup Environmental PIT Research Report: *Passive venting of soil gases beneath buildings*, 1997 (in two volumes – Volume 1, 60 pages and Volume 2, 56 pages). Passive gas protective systems are described in documents produced by BRE (especially BRE Report 414, see below) and CIRIA (see above). However, the information in these documents is of a practical nature and is essentially limited to guidance on general principles of use. This report provides quantitative information on the relative performance of various ventilation media and guidance on the design of passive gas protective measures. This report is highly technical and (unlike BRE 414, for example) contains very little practical guidance (e.g. Volume 2 contains example outputs of computational fluid dynamics modelling). Topics covered include:
  - site investigation requirements
  - factors influencing ventilation performance
  - example gas regimes studied
  - design considerations for ventilation layer
  - open voids
  - expanded polystyrene shuttering
  - geocomposite drainage blankets
  - granular blankets
  - granular blankets with drain networks
  - membranes
  - summary of gas dispersal characteristics of ventilation media.

The report is intended to be used by consultants, engineers and contractors who are engaged in the design of buildings on or near low gas potential sites. The guide is also intended to provide a reference document to regulators, such as Local Authority Building Control and Environmental Health Officers, to assist them in their assessment of ventilation schemes for particular developments.

The above discussion has mainly been in the context of housing development. In the context of non-domestic development the focus might be on the building only, but the general approach is the same.

4.6.5 Building near landfill sites and on gas contaminated land – gas-control measures

The investigations into gas risks carried out in accordance with the advice given above may conclude that the risks posed by the gases which are present are unacceptable. If this is the case, then appropriate remedial measures may be required to manage the risks. These can be applied to the building alone or, where the risks on any land associated with the building are deemed unacceptable, this could even mean adopting site-wide gas control measures which could include:

- removal of the material generating the gas, or
- covering together with systems for gas extraction.

CIRIA Report 149 (referred to above) or BRE Report 414 (referred to below) contains further guidance on this. In general, where site-wide gas-control measures are thought to be needed then expert advice should be sought.

Gas-control measures – dwellings

Practical guidance on construction methods to prevent the ingress of landfill gas in buildings is not given in Approved Document C. Instead, the reader is referred to BRE/Environment Agency Report BR 414 *Protective measures for housing on gas-contaminated land*, 2001 (70 pages) where detailed practical guidance on the construction of protective measures may be found. BRE Report 414 effectively replaces BRE Report 212 which was referred to in previous editions of AD C but is now referred to no longer.

Gas-control measures for dwellings are normally passive (i.e. the gas flow is driven by temperature differences (stack effect) and the effects of wind) and consist of a barrier which is gas resistant, across the entire walls and floor of the dwelling. Below this will be an extraction (or ventilation) layer from which gases can be dispersed and vented to atmosphere. In order to maximize the driving forces of natural ventilation it will be necessary to consider the design and layout of buildings.

Figure 4.6 gives typical examples of some of the constructional details contained in the BR 414.

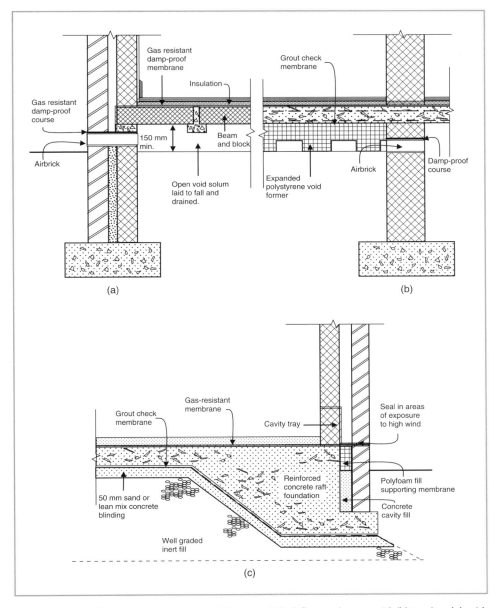

Figure 4.6 Landfill gas protection details: (a) Beam and block floor with open void; (b) *In situ* slab with EPS void former; (c) Raft foundation with membrane on top of slab

Gas-control measures – non-domestic buildings

In non-domestic buildings gas control measures are based on the same principles as those used for housing, therefore the DETR/Arup Environmental report referred to in the previous paragraph can also be used as a design guide. Since the floor areas of non-domestic buildings can be considerably larger than those of dwellings and the adequate dispersal of gas from beneath the floor must be ensured, it is usually necessary to seek expert advice. With larger floor areas passive systems may not be efficient at ensuring removal of gases, therefore mechanical systems (which may include monitoring and alarm systems) may be necessary. Such systems also need to be calibrated and continually maintained so they are more appropriate for non-domestic buildings where there is scope for this. Since special sub-floor ventilation systems are carefully designed to ensure adequate performance, they should not be modified unless a specialist review of the design is undertaken. It should be noted that the use of continuous mechanical ventilation for the removal of landfill gases in dwellings is not recommended since there is a risk of interference by users and maintenance of the system cannot be guaranteed. Consequently, a failure might result in a sharp increase in indoor methane concentration with the possibility of an explosion occurring.

It should be noted that the above brief notes on BR 414 are intended to give an idea of the content of that document. Designers of buildings which are likely to be affected by landfill gas should consult the full report and any of the references mentioned in sections 4.6.1 to 4.6.5 above.

Subsoil drainage

5.1 Provision of subsoil drainage

Subsoil drainage must be provided *if* it is necessary to avoid:

- the passage of moisture from the ground to the inside of the building, or
- damage to the fabric of the building. This includes damage to the foundations of the building caused by the transport of waterborne contaminants.

The provisions in AD C assume either that the site of the building will not be subject to general flooding or, if it is, then suitable steps are being taken. Interestingly, although flood resistance is not covered by the *Building Regulations 2000* at present, there is a presumption in planning guidance that development should not take place in areas that are at risk of flooding (see Planning Policy Guidance Note PPG 25 *Development and flood risk*, DTLR, 2002). This 59-page planning guidance document explains how flood risk should be considered at all stages of the planning and development process in order to reduce future damage to property and loss of life. The planning system seeks where possible to reduce and certainly not to increase flood risk. It should help ensure that flood plains are used for their natural purposes, continue to function effectively and are protected from inappropriate development. The guidance also outlines how flood risk issues should be addressed in regional planning guidance, development plans and in the consideration of planning applications.

5.2 Building in areas prone to flooding

Where local considerations might necessitate building in areas which are prone to flooding, the following guidance is offered to mitigate some of its effects:

- The existence of elevated groundwater levels or the flow of subsoil water across a site may be alleviated by the provision of adequate subsoil drainage (see below).
- The creation of blockages in drains and sewers caused by flooding can lead to backflow of sewage into buildings through low-sited toilets and gullies etc. This can be mitigated by the use of anti-flooding devices and non-return valves (see the CIRIA publication C506 *Low cost options for prevention of flooding from sewers*, 1998, 105 pages). This guidance document summarizes the results of a CIRIA project in which low-cost options for preventing flooding from sewers were identified and information collected on their suitability and effectiveness. Anti-flooding devices are dealt with in Appendix A1 of the publication. Different types are described together with the extent of their use, their cost and their reliability and effectiveness. The document is aimed at the needs of drainage engineers and planners working for sewerage undertakers, local authorities, contractors and developers.
- In areas where the design of the below-ground drainage system is such that foul water drainage also receives rainwater, these systems may surcharge in periods of heavy rainfall. This could lead to increased risks of localized flooding within properties in low-lying areas or in those which contain basements unless preventative measures are taken. Some guidance on protection is given in Approved Document H *Drainage and waste disposal* which may be summarized as follows:
    - Under conditions of heavy rainfall, combined and rainwater sewers are designed to surcharge, whereby the water level in the manhole rises above the top of the pipe. This may also happen to some foul sewers if they receive rainwater. Therefore, on some low-lying sites properties may be at increased risk of flooding if the ground level of the site (or the level of a basement) is below the level at which the drainage connects to the public sewer. The sewerage undertaker should be consulted in such cases to determine the extent and frequency of the likely surcharge.
    - Where a basement contains sanitary appliances and the sewerage undertaker considers that the risk of flooding due to surcharging is high, the drainage from the basement should be pumped. For low risks, an anti-flooding valve should be installed on the drainage from the basement.
    - For low-lying sites (i.e. those not containing basements) where the risk is low, protection for the building may be achieved by the provision of an

external gully sited at least 75 mm below floor level in a position so that any flooding from the gully will not damage any buildings. Higher risk areas should have anti-flooding valves or pumped drainage systems. Anti-flooding valves should:
- be of the double valve type
- be suitable for foul water
- have a manual closure device
- comply with prEN 13564 *Anti flooding devices for buildings*.

Normally, a single valve should serve only one building and information about the valve should be provided on a notice inside the building. The notice should indicate the location of any manual override and include necessary maintenance information. Some parts of the drainage system may be unaffected by surcharging. These parts should bypass any protective measures and should discharge by gravity.
- The passage of groundwater through a floor can be dealt with using water resistant construction (see section 6.3 below).
- The entry of water into underfloor voids can be addressed by making provision for the inspection and clearing out of such locations beneath suspended floors.

The following publications may also be consulted for further guidance on flooding and flood resistant construction:

- *Preparing for floods: interim guidance for improving the flood resistance of domestic and small business properties*, DTLR, 2002 (see section 4.4 above).
- *BRE for Scottish Office Design guidance on flood damage for dwellings*, TSO, 1996 (see section 4.4 above).
- *CIRIA/Environment Agency Flood products. Using flood protection products – a guide for home owners*, 2003 (12 pages). This handy guide is aimed specifically at home owners (and tenants, presumably) and is designed to help its target audience to assess the risk that flooding entails and to understand the different routes through which floodwater may enter. It explains what protection can be expected from flood protection products and other measures that householders can take to reduce floodwater from entering their property. Although the publication has been produced for homeowners with DIY or other technical experience, the information may also be useful for owners of small business properties. To accompany the guide, CIRIA has produced a series of advice sheets with more technical information on improving the flood resistance of buildings. These advice sheets and other up-to-date information on reducing the impacts of flooding can be downloaded from CIRIA's web site www.ciria.org/flooding.

5.3 Problems caused by subsoil water

Subsoil water may cause problems where:

- there is a high water table (i.e. within 0.25 m of the lowest floor in the building)
- surface water may enter or adversely affect the building
- an active subsoil drain is severed during excavations
- the stability and properties of the ground are adversely affected by groundwater beneath or around the building
- groundwater flows are altered by general excavation work for foundations and services.

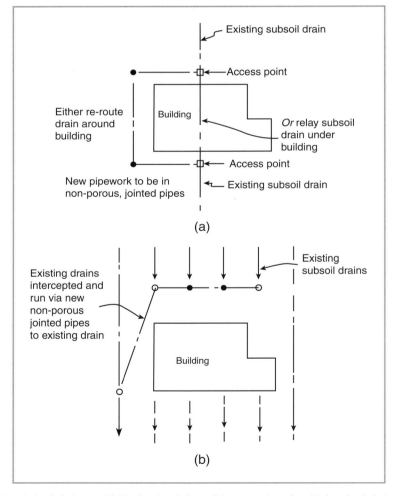

Figure 5.1 Subsoil drainage: (a) Single subsoil drain; (b) Interception of multiple subsoil drains

Where problems are anticipated it will usually be necessary either to drain the site of the building or to design and construct it to resist moisture penetration. Subsoil drainage should also be considered where contaminants are present in the ground, in order to prevent the transportation of waterborne contaminants to the foundations or into the building or its services.

Severed subsoil drains which pass under the building should be intercepted and continued in such a way that moisture is not directed into the building. Figure 5.1 illustrates a number of possible solutions.

PART 3

Resistance to moisture

Floors

6.1 Regulation requirements

Requirement C2 of Schedule 1 to the *Building Regulations 2000* deals with resistance to moisture. It requires the floors, walls and roof of a building to adequately protect it and the people who use it from harmful effects caused by:

- moisture from the ground
- precipitation and wind-driven spray
- surface and interstitial condensation
- water spilt from or associated with sanitary fittings and fixed appliances.

Guidance on the application of Requirement C2 to floors in Approved Document C is given in Section 4. It covers:

(a) ground-supported floors, suspended timber ground floors and suspended concrete ground floors when these are exposed to ground moisture
(b) interstitial condensation risk in ground floors and upper floors exposed from below
(c) surface condensation risk and mould growth on any floor type.

6.2 Definitions

In Approved Document C the following definitions apply with regard to Requirement C2:

FLOOR – the lower horizontal surface of any space in a building including any surface finish which is laid as part of the permanent construction. This would, presumably, exclude carpets, linoleum, tiles, etc., but would include screeds and granolithic finishes.

GROUNDWATER – liquid water (i.e. not water vapour, ice or snow etc.), either flowing through the ground or as a static water table.

INTERSTITIAL CONDENSATION – water vapour being deposited as liquid water within or between the layers of the envelope of the building.

MOISTURE – water present as a liquid, gas (e.g. water vapour) or solid (e.g. ice or snow etc.).

PRECIPITATION – moisture in any form falling from the atmosphere, such as rain, sleet, snow or hail, etc.

ROOF – any part of the external envelope of a building that makes an angle of less than 70° to the horizontal.

SPRAY – wind-driven droplets of water blown from the surface of the sea or other bodies of water close to buildings. The salt content of sea spray makes it especially hazardous to many building materials.

SURFACE CONDENSATION – water vapour being deposited as liquid water on visible surfaces within the building.

VAPOUR CONTROL LAYER – typically, this is a membrane material which is used in the construction, with the purpose of substantially reducing the transfer of water vapour through any building in which it is incorporated.

WALL – any opaque part of the external envelope of a building that makes an angle of 70° or more to the horizontal.

6.3 Protection of floors next to the ground

Ground floors should be designed and constructed so that:

• the passage of moisture to the upper surface of the floor is resisted (this might not apply to buildings used solely for storage of goods in which the only persons habitually employed were store people, etc. engaged only in taking in, caring for, or taking out the goods. Other similar types of buildings where the air is so moisture laden that any increase would not adversely affect the health and safety of the occupants might also be excluded)
• they will not be adversely affected by moisture from the ground
• they will not be adversely affected by groundwater
• the passage of ground gases is resisted. This relates back to Requirement C1 (2) (see section 4.1 above) where floors in certain localities may need to be constructed to resist the passage of hazardous gases such as methane or radon. The remedial measures shown in sections 4.5.2 to 4.6.5 above can function as both a gas resistant barrier and damp proof membrane if properly detailed

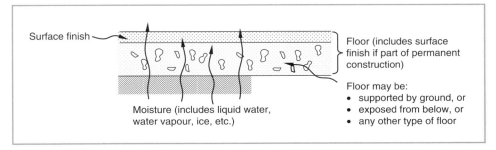

Figure 6.1 Floors – general guidance

- the structural and thermal performance of the floor is not adversely affected by interstitial condensation. This also applies to floors exposed from below
- surface condensation and mould growth is not promoted under reasonable occupancy conditions. This applies to all floors (not just those next to the ground).

This guidance is illustrated in Figure 6.1.

6.3.1 Floors supported directly by the ground

The requirements mentioned above can be met, for ground-supported floors, by covering the ground with dense concrete incorporating a damp-proof membrane, laid on a hardcore bed. If required, insulation may also be incorporated in the floor construction.

This form of construction is illustrated in Figure 6.2 and unless the floor is subjected to water pressure, such as occurs with buildings on very permeable strata like chalk, gravel or limestone (see note on the use of BS 8102 below), the construction of the floor should take into account the following points:

- Well-compacted hardcore less than 600 mm thick laid under the floor next to the ground should not contain water-soluble sulphates or deleterious matter in such quantities as might cause damage to the concrete. In situations where the use of a ground-supported floor would require a hardcore bed deeper than 600 mm there is a danger that excessive settlement might occur leading to cracking of the floor slab. Under such conditions a suspended floor construction would be a better alternative. Broken brick or stone is the best hardcore material. Clinker is dangerous unless it can be shown that the actual material proposed is free from sulphates etc., and colliery shales should likewise be avoided. In any event, the builder might well be liable for breach of an implied common law warranty of

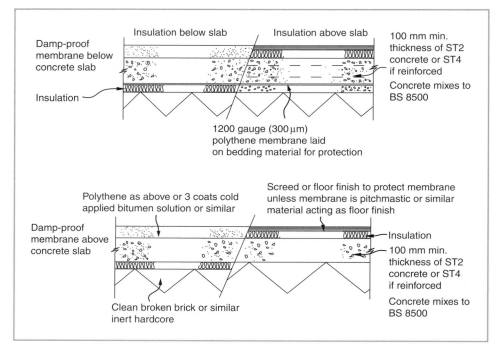

Figure 6.2 Ground-supported floor

fitness of materials (see *Hancock* v. *B. W. Brazier (Anerley) Ltd* [1966] 2 All ER 901), where builders were held liable for subsequent damage caused by the use of hardcore containing sulphates. Information on the choice of hardcore materials may be found in BRE Digest 276 *Hardcore*, 1992 (4 pages). This short, practical paper provides information on hardcore, considers the suitability of some of the materials in common use and discusses some of the problems that can arise. It deals with water-soluble sulphates, swelling due to chemical and volume changes in the material, consolidation of hardcore and materials used for hardcore. Like most other BRE Digests, this one is of universal application to all concerned with the built environment.

- A damp-proof membrane (DPM) may be provided above or below the concrete floor slab and should be laid continuous with the damp-proof courses in walls, piers, etc. If laid below the concrete, the DPM should be at least equivalent to 300 μm (1200 gauge) polyethylene (e.g. polythene). It should have sealed joints and should be supported by a layer of material that will not cause damage to the polythene. If a polyethylene sheet membrane is laid above the concrete, there is no need to provide the bedding material. It is also possible to use a three-coat layer of cold applied bitumen emulsion (or equivalent material with similar moisture and water vapour resistance) in this position. These materials should be protected by a suitable floor finish or screed. Surface protection does not need to be provided where the membrane consists of pitchmastic or similar material which also serves as a floor finish. There is one particular case where the membrane should be placed below the concrete slab, and that is where the ground contains water-soluble sulphates or other deleterious material that could contaminate the hardcore (see also BRE Special Digest SD1 *Concrete in aggressive ground*, 2003 in section 4.1.2 above).

- The minimum thickness for the concrete slab is 100 mm (although the structural design for the slab may require it to be thicker) designed to mix ST2 in BS 8500. Reinforced concrete should be to mix ST4 in BS 8500. BS 8500 (full title – BS 8500-1: 2002 *Concrete*. Complementary British Standard to BS EN 206-1. *Method of specifying and guidance for the specifier* (52 pages)) is essential reading for all people concerned with the specification and ordering of concrete for use on site. Approved Document C recommends that prescribed concrete mixes be used (ST1, ST2, ST4, etc.); however, BS 8500 actually offers the specifier five approaches to the specification of concrete as follows:
    - designated concretes (see Section 4.2 of the BS – see following bullet point)
    - designed concretes (see Section 4.3 of the BS – more flexibility is offered to the specifier than with designated concretes, which does not cover every application and every constituent material)
    - prescribed concretes (see Section 4.4 of the BS – allows the specifier to prescribe the exact composition and constituents of the concrete)
    - standardized prescribed concretes (see Section 4.5 of the BS – appropriate where concrete is site-batched on a small site or obtained from a ready-mixed concrete producer who does not have accredited third-party certification)
    - proprietary concretes (see Section 4.6 of the BS – appropriate where it is required that the concrete achieves a performance, using defined test methods, outside the normal performance requirements for concrete.
- Insulants located below floor slabs should:
    - have sufficient strength to resist the weight of the slab
    - be able to carry the anticipated floor loading
    - be able to support any overloading during construction
    - if placed below the damp-proof membrane have low water absorption (in order to resist degradation) and, if necessary, be resistant to contaminants in the ground.
- Approved Document C Section 4 gives no guidance on the position of the floor relative to outside ground level. Since this type of floor is unsuitable if subjected to water pressure, it is reasonable to assume that the top surface of the slab should not be below outside ground level unless special precautions are taken.

If it is proposed to lay a timber floor finish directly on the concrete slab, it is permissible to bed the timber in a material that would also serve as a damp-proof membrane. No guidance is given regarding suitable DPM materials. However 12.5 mm of asphalt or pitchmastic will usually be satisfactory for most timber finishes and it may be possible to lay wood blocks in a suitable adhesive DPM. If a timber floor finish is fixed to wooden fillets embedded in the concrete, the fillets should be treated with a suitable preservative unless they are above the DPM – see BS 1282: 1999 *Wood preservatives Guidance on choice, use and application* (16 pages). This British Standard provides an overview of wood preservation and the factors for consideration in the selection, and application of appropriate wood preservatives and in the use of preservative-treated timber. It also shows how the many British and European standards relating to specific aspects of wood preservation interrelate, and indicates which standards have further bearing on the topics raised. It covers:

- biological hazards (staining fungi, mould fungi, wood rotting fungi and wood boring beetles)
- natural durability
- the need for preservation
- types of preservatives and their characteristics (tar oil preservatives, waterborne preservatives and organic solvent preservatives)
- safe handling of preservatives
- assessing preservative efficacy
- the quantity of preservative required
- application of preservatives (industrial pre-treatment, preparation of wood prior to treatment, application techniques and *in situ* treatment)
- characteristics, handling and disposal of treated wood (drying, strength, compatibility with other materials, combustibility of treated wood, protection of wood treated with boron or light organic solvent preservative, machining and sanding treated wood and safety.

This is a non-technical standard and will be of use to anyone concerned with the treatment of wood in the built environment such as architects, building surveyors, specialist timber treatment companies and contractors.

Clause 11 of CP 102: 1973 *Protection of buildings against water from the ground* (34 pages) may be used as an alternative to the above. This code deals with the methods of preventing the entry of groundwater and surface water into a building from the surrounding areas. It makes recommendations for the drainage of adjoining areas, for special waterproof or water-resisting types of construction below ground level and for the damp-proofing of walls and floors at or near ground level. Clause 11 deals specifically with damp-proofing of floors where there may be capillary rise of moisture but not with those where water can exert a hydrostatic pressure. It covers:

- principles of damp-proofing
- factors influencing the degree of protection (such as temperature gradient, presence of aggregates containing alkali salts, degree of impermeability, durability)
- continuity of membrane with the damp-proof course in the surrounding walls
- preparation of site (need for subsoil drainage, support for the membrane, embedment of timber)
- work on site for waterproof flooring materials (such as mastic asphalt and pitchmastic flooring)
- work on site for sandwich membranes (mastic asphalt, bitumen damp-proof courses, hot laid pitch or hot bitumen, cold bitumen solution)
- damp-proofing for suspended timber floors.

Although limited due to its age (there are many new materials that are not covered by this code) the principles described remain sound and will be of particular use to designers, surveyors and contractors engaging in renovation and repair work. The code is well illustrated and easy to read, and will be of use to students.

Where groundwater pressure is evident, recommendations may be found in BS8102: 1990 *Code of practice for protection of structures against water from the ground*. This 40-page code of practice provides guidance on methods of dealing with and preventing the entry of water from surrounding ground into a building below ground level. The main methods described are the use of applied water-proofing finishes, watertight construction and drained cavity construction. The code does not cover the use of embedded heating in basements, floors and walls or for the special requirements in connection with the design and construction of cold stores. The code contains a useful amount of practical information on the waterproofing of buildings against ground moisture and will be of use to designers, general and specialist contractors, and regulatory authorities.

Again, the recommendations of Clause 11 of CP 102: 1973 may be used instead of the above, especially if the floor has a highly vapour-resistant finish.

6.3.2 Suspended timber ground floors

The performance requirements mentioned above may be met for suspended timber ground floors by:

- covering the ground with suitable material to resist moisture and deter plant growth

- providing a ventilated space between the top surface of the ground covering and the timber
- isolating timber from moisture-carrying materials by means of damp-proof courses.

A suitable form of construction is shown in Figure 6.3 and is summarized below:

- The ground surface should be covered with at least 100 mm of concrete to BS 8500 mix ST1 (see section 6.3.1 above for details of BS 8500), if unreinforced. It should be laid on clean broken brick or similar inert hardcore not containing harmful quantities of water-soluble sulphates or other materials which might damage the concrete. (The Building Research Establishment suggests that over 0.5% of water-soluble sulphates would be a harmful quantity.) Alternatively, the ground surface may be covered with at least 50 mm of concrete, as described above, or inert fine aggregate, laid on a polythene DPM as described for ground-supported floors above. The joints should be sealed and the membrane should be laid on a protective bed such as sand blinding.

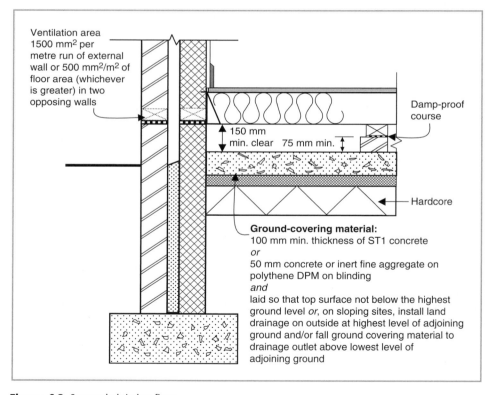

Ventilation area 1500 mm² per metre run of external wall or 500 mm²/m² of floor area (whichever is greater) in two opposing walls

Damp-proof course

150 mm min. clear   75 mm min.

Hardcore

Ground-covering material:
100 mm min. thickness of ST1 concrete
or
50 mm concrete or inert fine aggregate on polythene DPM on blinding
and
laid so that top surface not below the highest ground level *or*, on sloping sites, install land drainage on outside at highest level of adjoining ground and/or fall ground covering material to drainage outlet above lowest level of adjoining ground

Figure 6.3 Suspended timber floor

- Since it undesirable for water to collect on top of the ground-covering material under a timber floor, the ground-covering material should be laid so that *either* its top surface is not below the highest level of the ground adjoining the building *or*, where the site slopes, it might be necessary to install land drainage on the outside at the highest level of the ground adjoining the building and/or fall the ground-covering material to a drainage outlet above the lowest level of the adjoining ground.

- There should be a space above the top of the concrete of at least 75 mm to any wall-plate and 150 mm to any suspended timber (or insulation where provided). This depth may need to be increased where the building is constructed on shrinkable clays in order to allow for heave.

- There should be ventilation openings in two opposing external walls allowing free ventilation to all parts of the sub-floor. An actual ventilation area equivalent to 1500 mm^2 per metre run of external wall or 500 mm^2/m^2 of floor area (whichever area is greater) should be provided and any ducts needed to convey ventilating air should be at least 100 mm in diameter. Ventilation openings should be fitted with grilles so as to prevent vermin entry but these grilles should not unduly resist the flow of air. It may be difficult, where there is a requirement for level access to the floor, to provide the ventilators in the position shown in Figure 6.3 since the top surface of the floor may well be nearer to the ground. The problem can usually be solved using offset (periscope) ventilators.

- Damp-proof courses of impervious sheet materials, slates or engineering bricks bedded in cement mortar should be provided between timber members and supporting structures to prevent transmission of moisture from the ground. BS 5628: *Code of practice for use of masonry.* Part 3: 2001 *Materials and components, design and workmanship* (132 pages) gives guidance on the choice of suitable materials. Part 3 of the BS gives general recommendations for the design, construction and workmanship of masonry, including materials and components and the main aspects of design. Information on damp-proof courses is contained in Section 5.5.5 of the BS, the performance of individual materials being given in its Table 3. Information on workmanship is given in Annex A, Section A5 of the BS. This Standard is essential reading for all people concerned with the specification and detailed design of buildings.

- In areas where water may be spilled (such as bathrooms, utility rooms and kitchens) boards used for flooring should be moisture resistant, irrespective of the storey in which they are located. Softwood boarding should be:
  - a minimum of 20 mm thick, and either
  - from a durable species (see BRE Digest 429 *Timbers and their natural durability and resistance to preservative treatment* (8 pages) where an

explanation of the classification of durability and treatability for timber may be found. The Digest classifies these properties for over 150 species – see Table 3 from the Digest. It is concerned with natural durability only in relation to fungal decay and not to resistance to attack by insects and marine borers), or

○ treated with a suitable preservative.

Chipboard is particularly susceptible to moisture damage, so where this is used as a flooring material it should be:

○ of one of the grades recommended as having improved moisture resistance specified in:

  – BS 7331:1990 *Specification for direct surfaced wood chipboard based on thermosetting resins* (16 pages). The standard specifies the requirements for six types of direct surfaced wood chipboard for interior applications as follows:

    ○ in high humidity conditions
    ○ in low and normal humidity conditions

    The required grades may be found in Table 1 of the BS as DH/C3 (high wear resistance and improved moisture resistance), DH/C5 (high wear resistance, improved moisture resistance and enhanced mechanical properties), DG/C3 (general purpose and improved moisture resistance) and DG/C5 (general purpose, improved moisture resistance and enhanced mechanical properties). The manufacturer is required to mark each batch of boards or panels on the package or on the material itself with the following information:

    ○ number and date of the British Standard (i.e. BS 7331:19901)
    ○ classification
    ○ manufacturer
    ○ batch number, or

  – BS EN 312 Part 5 *Particleboards. Specifications. Requirements for load-bearing boards for use in humid conditions*: 1997 (16 pages). Particleboards are required to comply with Tables 1 and 2 of the Standard. Table 1 relates to the mechanical and swelling properties of the boards (bending strength, modulus of elasticity in bending, internal bond, swelling in thickness after 24 hours), whereas Table 2 contains requirements for moisture resistance (either internal bond after cyclic test and swelling in thickness after cyclic test, or internal bond after boil test). Each panel has to be clearly and indelibly marked by the manufacturer, with the following information:

    ○ the manufacturer's name, trademark, or identification mark
    ○ the number of the European Standard (i.e. EN 312-5)
    ○ the nominal thickness

○ the formaldehyde class

○ the batch number, or the production week and year.

Each of the above three references will be of particular use to people concerned with the specification and detailed design of timber floors.

o laid, fixed and jointed in the manner recommended by the manufacturer with the identification marks facing upwards in order to demonstrate compliance.

Again, the recommendations of Clause 11 of CP 102: 1973 (see section 6.3.1 above) may be used instead of the above, especially if the floor has a highly vapour-resistant finish.

6.3.3 Suspended concrete ground floors

Moisture should be prevented from reaching the upper surface of the floor and the reinforcement should be protected against moisture if the construction is to be considered satisfactory.

Suitable suspended concrete ground floors may be of precast construction with or without infilling slabs or they may be cast *in situ*.

Normally, for *in situ* construction the concrete should be at least 100 mm thick (unless required to be thicker by the structural design) and it should contain a minimum of 300 kg of cement per m^3 of concrete. The reinforcement should be protected by at least 40 mm cover.

Precast concrete construction offers another solution and this can be built with or without infilling slabs or blocks. The reinforcement in this case should have at least the thickness of cover required for moderate exposure.

A damp-proof membrane should be provided if the ground below the floor has been excavated so that it is lower than outside ground level and it is not effectively drained. The space between the underside of the floor and the ground should be ventilated. The space should be at least 150 mm in depth (measured from the ground surface to the underside of the floor or insulation, if provided) and the ventilation recommendations should be as for suspended timber floors (see paragraph 6.3.2).

If the building is located in an area where flooding might be a problem, it may be necessary to include a means of inspecting and clearing out the sub-floor voids beneath suspended floors. Further guidance on this can be obtained from the DTLR publication *Preparing for floods: interim guidance for improving the flood resistance of domestic and small business properties*, DTLR, 2002 (see section 4.4 above for details of this publication).

These recommendations are summarized in Figure 6.4.

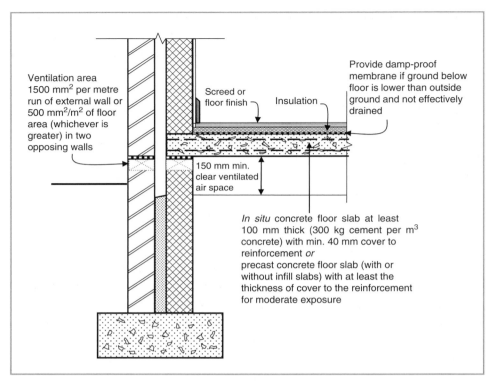

Ventilation area 1500 mm² per metre run of external wall or 500 mm²/m² of floor area (whichever is greater) in two opposing walls

Screed or floor finish

Insulation

Provide damp-proof membrane if ground below floor is lower than outside ground and not effectively drained

150 mm min. clear ventilated air space

In situ concrete floor slab at least 100 mm thick (300 kg cement per m³ concrete) with min. 40 mm cover to reinforcement *or*
precast concrete floor slab (with or without infill slabs) with at least the thickness of cover to the reinforcement for moderate exposure

Figure 6.4 Suspended concrete ground floors

6.3.4 Resistance to damage from interstitial condensation for ground floors and floors exposed from below

Ground floors and floors exposed from below (such as those above an open parking bay under a building or an open passageway), should be designed and constructed so that their structural and thermal performance are not adversely affected by interstitial condensation. No actual guidance is given in Approved Document C, the reader merely being referred to other sources of guidance such as:

• Clause 8.5 and Annex D of BS 5250: 2002 *Code of practice for the control of condensation in buildings*. This 82-page British Standard code of practice describes the causes and effects of surface and interstitial condensation in buildings and gives recommendations for their control. The principles of control and the recommendations given can be applied generally to both new and existing buildings, although some constructions (e.g. curtain walling or those around cold stores and those buildings with unusually high internal humidities) are outside the scope of the Standard since they would need specialized treatment.

The Standard includes recommendations for heating, ventilation and construction which can control condensation and gives methods of calculation to help assess and quantify risk. Methods are given to determine the occurrence and assess the effects of:

○ surface condensation, or mould growth (one of its associated effects)
○ interstitial condensation.

Clause 8.5 is specifically related to floors and covers the following main types:

○ suspended, where the floor structure spans a void or crawlspace
○ solid, where the floor rests directly on prepared ground
○ externally exposed floors, where the building shape results in the underside of a floor being exposed to the outside air.

Design considerations are described and design details are given for the three types of floors. Annex D contains a summary of the calculation methods for assessing the risk of surface condensation and mould growth; and interstitial condensation. BS EN ISO 13788 (see below) should be consulted for the full calculation methods. This is an easily readable Standard and will be of most use to building designers, contractors, owners, managers and occupiers.

• BS EN ISO 13788: 2001 *Hygrothermal performance of building components and building elements. Internal surface temperature to avoid critical surface humidity and interstitial condensation. Calculation methods.* This 40-page Standard gives calculation methods for:

  ○ The internal surface temperature of a building component or building element below which mould growth is likely, given the internal temperature and relative humidity – the method can also be used to assess the risk of other surface condensation problems.

  ○ The assessment of the risk of interstitial condensation due to water vapour diffusion. The method used assumes built-in water has dried out and does not take account of a number of important physical phenomena including:

    – the dependence of thermal conductivity on moisture content
    – the release and absorption of latent heat
    – the variation of material properties with moisture content
    – capillary suction and liquid moisture transfer within materials
    – air movement through cracks or within air spaces
    – the hygroscopic moisture capacity of materials.

  Consequently the method is applicable only to structures where these effects are negligible. This Standard is mainly aimed at specialists with the required level of knowledge of physics and higher mathematics and is not really for the general reader.

• BRE Report BR 262 *Thermal insulation: avoiding risks*, 2002 (85 pages). This excellent guide discusses the more important technical risks associated with meeting the requirements of building regulations for thermal insulation. Technical risks are highlighted and these are followed by actions that could be taken to avoid the risk. The report is divided into five sections relating to the major elements of the building. These sections are broken down into subsections to reflect alternative construction methods and the impact of the position of insulation within the construction. Each section concludes with a list of quality control checks for use on site. Illustrations outline generic construction principles and good practice details. The guide does not try to impose its own solutions on designers and builders since it recognizes that they may have established details that are equally suitable. Section 5 of the report deals specifically with floors under the following headings:

- Concrete ground floors insulated below the structure
- Concrete ground floors insulated above the structure
- Concrete ground floors insulated at the edge
- Suspended timber ground floors
- Concrete and timber upper floors
- Quality control checks for floors

As with most of the BRE's publications this report should form part of the reference library of all those concerned with the design, construction and use of buildings.

6.3.5 *Resistance to surface condensation and mould growth in floors*

In order to resist surface condensation and mould growth in floors it is necessary to ensure that the surface is maintained above the dewpoint temperature. This will depend on the outside temperature, the temperature of the room in which the floor surface is situated and the relative humidity of the room. It can be affected by ventilation, by thermal bridging of construction elements at junctions and by the degree of thermal insulation provided in the flooring elements.

Therefore, in all floors, care should be taken to design the junctions between the elements so that thermal bridging is avoided. This can be done by following:

- the recommendations in the report *Limiting thermal bridging and air leakage: robust construction details for dwellings and similar buildings*, published by The Stationery Office, 2002 (in 8 volumes). The Robust Details have been prepared to assist the construction industry in achieving the performance standards published in the Building Regulations Approved Documents L1A, L1B, L2A and L2B (2006 edition). They are intended to reduce risks and potential problems that can arise as a result of building to higher energy efficiency standards such as:

o interstitial condensation risk which can lead to the deterioration of the structure

o surface condensation risk which can lead to mould growth

o blockage of essential ventilation paths which can lead to condensation building up in places such as roof or floor voids

o risk of rain penetration

o higher heat loss than expected

o higher air leakage than expected – leading to unforeseen but significant heat losses and occupant discomfort.

There are certain advantages to be gained by using the Robust Details. For example, on new dwellings the requirement for air pressure testing is restricted to a single unit of each dwelling type in a development if the approved details are followed, whereas if they have not been adopted then a greater number of dwellings will need to be tested (the exact number being determined by the number of dwellings of each type in the development). The Robust Details cover constructions of the following general types (the parts that cover floor details are in brackets):

o Section 2. Masonry: External Wall Insulation (parts 2.11 to 2.20)

o Section 3. Masonry: Cavity Wall Insulation: Full-Fill (parts 3.13 to 3.22)

o Section 4. Masonry: Cavity Wall Insulation: Partial-Fill (parts 4.13 to 4.22)

o Section 5. Masonry: Internal Wall Insulation (parts 5.12 to 5.20)

o Section 6. Timber Frame (parts 6.11 to 6.18)

o Section 7. Light Steel Frame (parts 7.07 to 7.13).

The details have been rigorously analysed to confirm they are robust, if constructed with reasonable attention to workmanship and supervision. Therefore they will be of great practical use to both designers and contractors and are essential reading, or

• the guidance of BRE Information Paper IP17/01 *Assessing the effects of thermal bridging at junctions and around openings* (8 pages). This information paper gives guidance on assessing the effects of thermal bridging at junctions and around openings in the external elements of buildings and how to assess their effect on the overall heat loss. It assumes that the reader has an understanding of the principles, and familiarity with the calculation, of the fabric heat loss through the plane external elements of buildings. The paper deals specifically with non-repeating thermal bridges (such as junctions of floor and roof with the external wall and details around window and door openings) where the additional heat flow due to the presence of this type of thermal bridge is determined separately. It gives the requirements for limiting the risk of surface condensation or mould growth at these thermal bridges and describes how to assess their thermal performance and how to incorporate the additional heat loss through such thermal bridges with that through the building fabric as a whole. Although the mathematics required to carry out the essential calculations may seem daunting at first, three example calculations are given which demonstrate the processes involved. Unfortunately, these are based on design methods (elemental, target U-value and trade-off) from the now obsolete 2002 edition of Approved Document L. Therefore, these methods are no longer deemed to be acceptable ways of satisfying the regulation requirements for conservation of fuel and power. It would seem that the information paper will only be of use to specialists dealing with energy use in buildings.

Additionally, ground floors should be designed and constructed so that the thermal transmittance (U-value) does not exceed 0.7 W/m^{2}K at any point.

7

Walls

7.1 Regulation requirements

Requirement C2 of Schedule 1 to the *Building Regulations 2000* deals with resistance to moisture. It requires the floors, walls and roof of a building to adequately protect it and the people who use it from harmful effects caused by:

- moisture from the ground
- precipitation and wind-driven spray
- surface and interstitial condensation
- water spilt from or associated with sanitary fittings and fixed appliances.

Guidance on the application of Requirement C2 to walls is given in Approved Document C in Section 5.

7.2 Definitions

In Approved Document C the following definitions apply with regard to Requirement C2:

FLOOR – the lower horizontal surface of any space in a building including any surface finish which is laid as part of the permanent construction. This would, presumably, exclude carpets, linoleum, tiles, etc., but would include screeds and granolithic finishes.

GROUNDWATER – liquid water (i.e. not water vapour, ice or snow, etc.), either flowing through the ground or as a static water table.

INTERSTITIAL CONDENSATION – water vapour being deposited as liquid water within or between the layers of the envelope of the building.

MOISTURE – water present as a liquid, gas (e.g. water vapour) or solid (e.g. ice or snow etc.).

PRECIPITATION – moisture in any form falling from the atmosphere, such as rain, sleet, snow or hail, etc.

ROOF – any part of the external envelope of a building that makes an angle of less than 70° to the horizontal.

SPRAY – wind-driven droplets of water blown from the surface of the sea or other bodies of water close to buildings. The salt content of sea spray makes it especially hazardous to many building materials.

SURFACE CONDENSATION – water vapour being deposited as liquid water on visible surfaces within the building.

VAPOUR CONTROL LAYER – typically, this is a membrane material which is used in the construction, with the purpose of substantially reducing the transfer of water vapour through any building in which it is incorporated.

WALL – any opaque part of the external envelope of a building that makes an angle of 70° or more to the horizontal.

7.3 Protection of walls against moisture from the ground

The term 'wall' means vertical construction, which includes piers, columns and parapets and may include chimneys if they are attached to the building. Windows, doors and other openings are not included, but the joint between the wall and the opening is included.

Walls should be constructed so that:

- the passage of moisture from the ground to the inside of the building is resisted (this might not apply to buildings used solely for storage of goods in which the only persons habitually employed were storemen, etc. engaged only in taking in, caring for or taking out the goods. Other similar types of buildings where the air is so moisture-laden that any increase would not adversely affect the health and safety of the occupants might also be excluded)
- they will not be adversely affected by moisture from the ground
- they will not transmit moisture from the ground to another part of the building that might be damaged.

The requirements mentioned above can be met for internal and external walls by providing a damp-proof course of suitable materials in the required position.

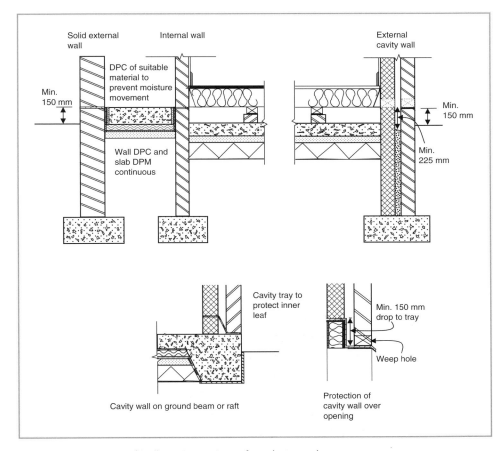

Figure 7.1 Protection of walls against moisture from the ground

Figure 7.1 illustrates the main provisions, which are summarized below:

- The damp-proof course may be of any material that will prevent moisture movement. This would include bituminous sheet materials, engineering bricks or slates laid in cement mortar, polyethylene or pitch polymer materials.
- The damp-proof course and any damp-proof membrane in the floor should be continuous.
- Unless an external wall is suitably protected by another part of the building, the damp-proof course should be at least 150 mm above the outside ground level.
- Where a damp-proof course is inserted in an external cavity wall, the cavity should extend at least 225 mm below the level of the lowest damp-proof course. Alternatively, precipitation (see definition in section 7.2 above) can be prevented

from reaching the inner leaf of the wall by the use of a damp-proof tray. This may be particularly useful where a cavity wall is built directly off a raft foundation, ground beam or similar supporting structure, and it is impractical to continue the cavity down 225 mm. Where a cavity tray is inserted, weep holes should provided every 900 mm in the outer leaf in order to allow moisture collecting on the tray to pass out of the wall. In some circumstances, such as above a window or door opening, the cavity tray will not extend along the full length of the exposed wall. Here stop ends should be provided to the tray and at least two weep holes should be provided.

For walls which are not subject to groundwater pressure, reference should be made to the relevant recommendations of Clauses 4 and 5 of BS 8215: *1991 Code of practice for design and installation of damp-proof courses in masonry construction* (20 pages). This British Standard contains recommendations for the selection, design and installation of damp-proof courses (DPCs) in both solid and cavity masonry constructions. Clauses 4 and 5 deal respectively, with materials and design. The Standard covers DPC materials in the following three groups:

- flexible materials (e.g. sheet lead, sheet copper, bitumen DPC, bitumen/sheet metal composites, polyethylene, bitumen polymer, pitch polymer
- semi-rigid materials (e.g. mastic asphalt)
- rigid materials (e.g. dense bricks and slates, bedded in cement mortar).

Any chosen DPC material should comply with the standards listed in Table 1 of BS 8215. Clause 5 – Design, provides information on the selection of the appropriate DPC material. Selection will depend on a number of factors, notably:

- exposure conditions of the site and the building
- provision of primary protection
- the degree of integration between the DPC and other parts of the construction
- the form of construction (e.g. solid or cavity masonry wall).

This Standard is essential reading for all people concerned with the specification and detailed design of buildings.

For walls subject to groundwater pressure, including basement walls, reference should be made to BS 8102: 1990 *Code of practice for protection of structures against water from the ground* (see section 6.3.1 above). This code of practice provides guidance on methods of dealing with and preventing the entry of water from surrounding ground into a building below ground level. Guidance on the design and construction of tanking systems is given in Section 3 of the code. The principle tanking methods described are mastic asphalt, bitumen sheet, internally applied cementitious waterproof render, polyurethane resin and self-adhesive rubber bitumen membrane. For each tanking method advice is given on:

- externally applied tanking
- internally applied tanking
- pumping
- site preparation
- thickness and finish, number of layers, etc.

The code contains a useful amount of practical information on the waterproofing of buildings against ground moisture and will be of use to designers, general and specialist contractors, and regulatory authorities.

7.4 Weather resistance of external walls

In addition to resisting ground moisture, external walls should:

- resist the passage of precipitation to the inside of the building (this might not apply to buildings used solely for storage of goods in which the only persons habitually employed were store people, etc. engaged only in taking in, caring for, or taking out the goods. Other similar types of buildings where the air is so moisture-laden that any increase would not adversely affect the health and safety of the occupants might also be excluded)
- not transmit moisture due to precipitation to other components of the building that might be damaged
- be designed and constructed so as not to allow interstitial condensation to adversely affect their structural and thermal performance
- not promote surface condensation and mould growth under reasonable occupancy conditions.

There are a number of forms of wall construction which will satisfy the above requirements:

- a solid wall of sufficient thickness holds moisture during bad weather until it can be released in the next dry spell
- an impervious or weatherproof cladding prevents moisture from penetrating the outside face of the wall
- the outside leaf of a cavity wall holds moisture in a similar manner to a solid wall, the cavity preventing any penetration to the inside leaf.

These principles are illustrated in Figure 7.2.

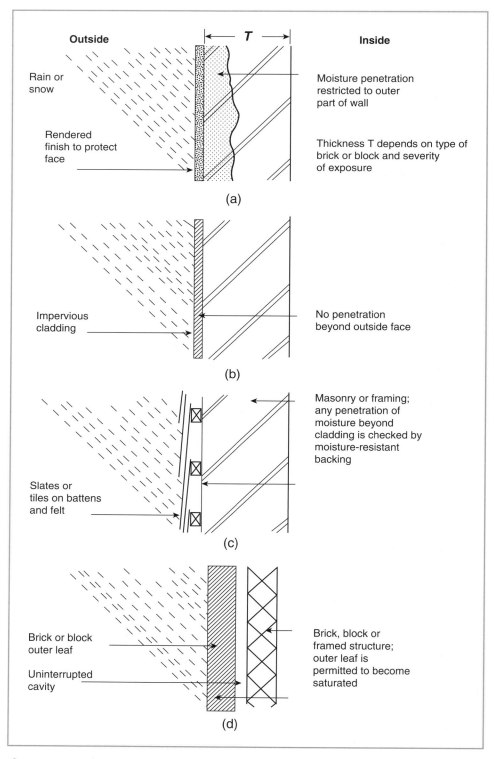

Figure 7.2 Weather resistance of external walls – principles: (a) Solid external wall; (b) Impervious cladding; (c) Weather-resistant cladding; (d) Cavity wall

7.4.1 Solid external walls

The thickness of a solid external wall will depend on the type of brick or block used and the severity of exposure to wind-driven rain. This may be assessed for a building in a given area by using BS 8104: 1992 *Code of practice for assessing exposure of walls to wind-driven rain* (75 pages). This British Standard gives recommendations for two methods for assessing exposure of walls in buildings to wind-driven rain (i.e. the local spell index method and the local annual index method). Although this code is not appropriate in respect of the weathertightness of components such as windows, the performance of such components can be influenced by the design of the surrounding structure and Appendix E from the code may be helpful in this respect. Features of this code of practice are that it allows calculations of driving rainfall for different orientations, it allows annual average values to be calculated as well as quantities for the worst likely spell in any three-year period and it allows corrections to be made for ground terrain, topography, local shelter, and the constructional characteristics of the building concerned. The code is relatively easy to use and contains in Appendix A, a number of worked examples using the wind-driven rain maps illustrated in Appendix B and topography factors in Appendix C. The code will mainly be of use to designers and specifiers when anticipating the possible use of solid wall construction in masonry.

Reference may also be made to BS 5628 *Code of practice for use of masonry* Part 3: 2001 *Materials and components, design and workmanship* (see section 6.3.2 above). Information on masonry construction is contained in Sections 5.6 and 5.11 of the code. Information on workmanship is given in Annex A, Section A5. This Standard is essential reading for all people concerned with the specification and detailed design of buildings.

In conditions of very severe exposure it may be necessary to use an external cladding. However, in conditions of severe exposure a solid wall may be constructed as shown in Figure 7.3. The following points should also be considered:

- For brickwork or stonework the wall should be at least 328 mm thick.
- For dense aggregate blockwork the wall should be at least 250 mm thick.
- For lightweight aggregate or aerated autoclaved concrete the wall should be at least 215 mm thick.
- The brickwork or blockwork should be rendered or given an equivalent form of protection.
- Rendering should have a scraped or textured finish and be at least 20 mm thick in two coats. This permits easier evaporation of moisture from the wall.
- The bricks or blocks and mortar should be matched for strength to prevent cracking of joints or bricks, and joints should be raked out to a depth of at least 10 mm in order to provide a key for the render.

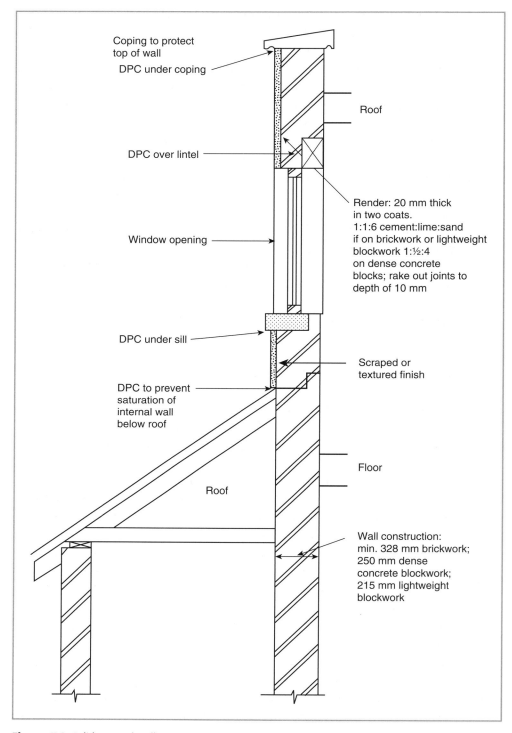

Figure 7.3 Solid external walls – moisture exclusion

Further guidance on mortar mixes is given in BS EN 998: *Specification of mortar for masonry* Part 2: 2002: *Masonry mortar* (24 pages). This part of BS EN 998 specifies requirements for masonry mortars (bedding, jointing and pointing) for use in masonry (e.g. facing and rendered masonry, load-bearing or non-load-bearing masonry structures, including internal linings and partitions, for building and civil engineering). It defines for fresh mortars the performance related to workable life, chloride content, air content, correction time (for thin layer mortars) and density. For hardened mortars it defines the performance related to compressive strength, bond strength, durability and thermal properties and density. All performances are measured according to the corresponding test methods contained in separate European Standards. The standard covers:

- materials (inorganic binders, aggregate, admixtures, additions and mixing water)
- requirements (properties of fresh mortar, workable life, chloride content, air content, properties of hardened mortar, additional requirements for thin layer mortars and mixing of mortar on site)
- designation and classification of masonry mortar
- marking and labelling
- evaluation of conformity.

There are two annexes covering, in Annex A, the sampling for initial type testing and independent testing of consignments, and in Annex B, the use of masonry units and masonry mortar. This Standard is essential reading for all people concerned with the specification and detailed design of buildings.

- The render mix should not be too strong or cracking may occur. A mix of 1:1:6 cement:lime:well graded sharp sand is recommended for all walls except those constructed of dense concrete blocks where 1:1/2:4 should prove satisfactory. Further details of a wide range of render mixes according to the severity of exposure and the type of brick or block may be obtained from BS 5262: 1991 *Code of practice for external renderings* (46 pages). This code of practice gives recommendations for cement-based external renderings on all common types of background. It includes renderings on both new and old backgrounds and the maintenance and repair of existing work. Renderings for liquid retaining structures such as tanks and manholes and renderings applied as a background for any form of cladding are not included. The code covers materials, design, backgrounds, work on site and maintenance and repair, and contains a series of tables giving constituents of mixes as follows:
  - Table 1. Mixes suitable for rendering
  - Table 2. Severe exposure: recommended rendering specifications
  - Table 3. Moderate and sheltered exposure: recommended rendering specifications

○ Table 4. Summary of precautions for rendering on to various types of new backgrounds

○ Table 5. Summary of recommendations for rendering on to contaminated or deteriorated existing backgrounds.

The code is easily readable and will be invaluable to anyone concerned with the specification and detailing of building work involving rendering.

- It is, of course, possible to obtain a wide range of premixed and proprietary mortars and renders. These should be used in accordance with the manufacturer's instructions.
- Where the top of a wall is unprotected by the building structure it should be protected to resist moisture from rain or snow. Unless the protection and joints form a complete barrier to moisture, a damp-proof course should also be provided.
- Damp-proof courses, trays and closers should be provided to direct moisture towards the outside face of the wall in the positions shown in Figure 7.3.
- Insulation to solid external walls may be provided on the inside or outside of the wall. Externally placed insulation should be protected unless it is able to offer resistance to moisture ingress so that the wall may remain reasonably dry (and the insulation value may not be reduced). Internal insulation should be separated from the wall construction by a cavity to give a break in the path for moisture. (Some examples of external wall insulation are given in Figure 7.4.)

7.4.2 External cavity walls

In order to meet the performance requirements, an external cavity wall should consist of an internal leaf which is separated from the external leaf by:

- a drained air space, or
- some other method of preventing precipitation from reaching the inner leaf.

An external cavity wall may consist of the following:

- An outside leaf of masonry (brick, block, natural or reconstructed stone).
- Minimum 50 mm uninterrupted cavity. Where a cavity is bridged (by a lintel, etc.) a damp-proof course or tray should be inserted in the wall so that the passage of moisture from the outer to the inner leaf is prevented. This is not necessary where the cavity is bridged by a wall tie, or where the bridging occurs, presumably, at the top of a wall and is then protected by the roof. Cavity walls may also be bridged by cavity barriers and fire stops where other parts of the Building Regulations require this (such as Part B or Part E). Where an opening

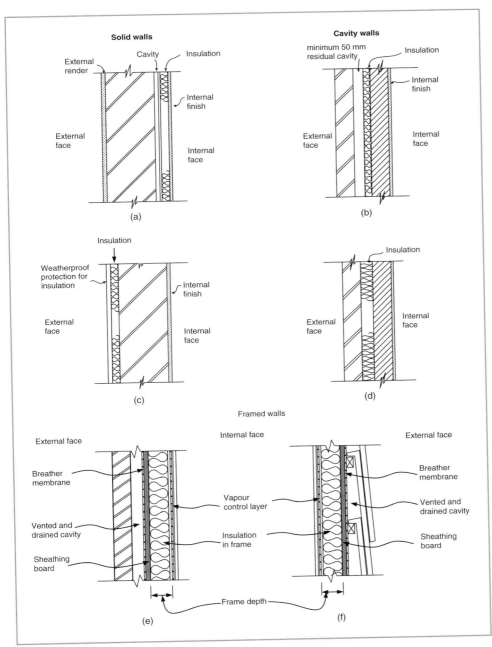

Figure 7.4 Weather resistance and insulation of external walls: (a) Internal insulation; (b) Partially filled cavity; (c) External insulation; (d) Fully-filled cavity; (e) Brick-clad timber-framed wall; (f) Tile-clad timber-framed wall

is formed in a cavity wall, the jambs should have a suitable vertical damp-proof course or the cavity should be provided with a suitable cavity closure so as to prevent the passage of moisture.

- An inside leaf of masonry or framing with suitable lining.
- In order to ensure structural robustness and weather resistance in the wall, masonry units should be laid on a full bed of mortar and the cross joints should be continuously and substantially filled.
- Where a cavity is only partially filled with insulation, the remaining cavity should be at least 50 mm wide (see Figure 7.4).

Alternatively, the relevant recommendations of BS 5628 *Code of practice for use of masonry*. Part 3: 2001 *Materials and components, design and workmanship* may be followed (see sections 6.3.2 and 7.4.1 above for details). Factors affecting rain penetration of cavity walls are indicated in the code in Clause 5.5.4.2.4.

7.4.3 Weather resistance and cavity insulation

Since the installation of cavity insulation effectively bridges the cavity of a cavity wall and could give rise to moisture penetration to the inner leaf, it is most important that it be carried out correctly and efficiently. Approved Document C lists a number of British Standards, codes of practice and other documents that cover the various materials that may be incorporated into a cavity wall. It also sets out a simple method for determining the minimum width of cavity, whether filled or clear, for various exposure conditions, various insulating methods and various forms of wall construction. Diagram 12 from Approved Document C is reproduced here in Figure 7.5. When used with Table 4 (reproduced here in Table 7.1) the suitability of a wall for installing insulation may be determined as follows:

(1) From Diagram 12 determine the national exposure (e.g. zone 1, 2, 3 or 4).
(2) Modify this result by either adding 1 to the exposure zone if local conditions accentuate wind effects (e.g. open hillsides or valleys where the wind is funnelled onto the wall), or subtract 1 from the exposure zone where the walls do not face into the prevailing wind.
(3) Select a form of wall construction from impervious cladding, rendered finish or facing masonry.
(4) Choose an insulation method from built-in full fill, injected fill (UF foam or other material), partial fill, internal insulation or fully filled cavity.
(5) Select the appropriate cavity width by reference to the exposure zone figure.

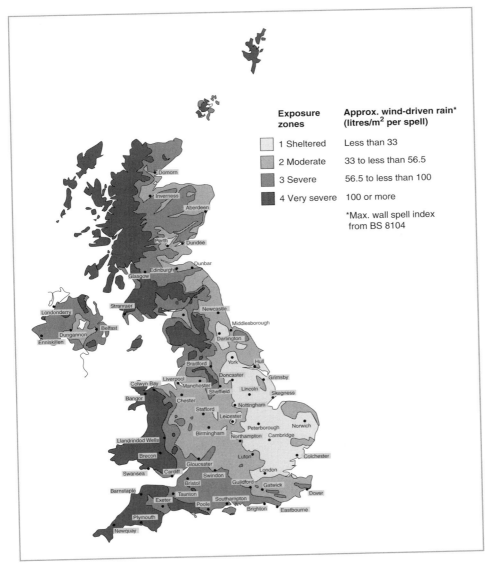

Figure 7.5 Approved Document C – Diagram 12: UK zones for exposure to driving rain

BRE Report 262 *Thermal insulation: avoiding risks* (see section 6.3.4 above) contains further information regarding the use of Table 4. Section 3 of the report deals specifically with walls under the following headings:

- Rain penetration through cavity and solid walls
- Masonry cavity walls
- Masonry walls with internal insulation

Table 7.1 Approved Document C, Table 4 – Maximum recommended exposure zones for insulated masonry walls

Wall construction	Min. width of filled or clear cavity (mm)	Maximum recommended exposure zone for each construction						
		Impervious cladding		Rendered finish		Facing masonry		
Insulation method		Full height of wall	Above facing masonry	Full height of wall	Above facing masonry	Tooled flush joints	Processed mortar joints	Flush sills and copings
Built-in full fill	50	4	3	3	3	2	1	1
	75	4	3	4	3	3	1	1
	100	4	4	4	3	3	1	2
	125	4	4	4	3	3	1	2
	150	4	4	4	4	4	1	2
Injected fill *not* UF foam	50	4	2	3	2	2	1	1
	75	4	3	4	3	3	1	1
	100	4	3	4	3	3	1	1
	125	4	4	4	3	3	1	2
	150	4	4	4	4	4	1	2
Injected fill UF foam	50	4	2	3	2	1	1	1
	75	4	2	3	2	2	1	1
	100	4	2	3	2	2	1	1
Partial fill								
Residual 50 mm cavity	50	4	4	4	4	3	1	1
Residual 75 mm cavity	75	4	4	4	4	4	1	1
Residual 100 mm cavity	100	4	4	4	4	4	2	1
Internal insulation								
Clear cavity 50 mm	50	4	3	4	3	3	1	1
Clear cavity 100 mm	100	4	4	4	4	4	2	2
Fully filled								
Cavity 50 mm	50	4	3	3	3	2	1	1
Cavity 100 mm	100	4	4	4	3	3	1	2

- Masonry walls with external insulation
- Walls of timber-framed construction
- Quality control checks for walls.

It is also possible to determine the suitability of a wall for installing insulation into the cavity by following the calculation or assessment procedure in current British or CEN standards. When partial fill materials are to be used, the residual cavity should not be less than 50 mm nominal.

The following points should also be taken into account:

- Rigid materials (boards or batts) which are built in as the wall is constructed should be the subject of current certification from an appropriate body or a European Technical Approval and the work should be carried out to meet the requirements of that document.
- Urea-formaldehyde foam inserted after the wall has been constructed should comply with BS5617: 1985 *Specification for urea-formaldehyde (UF) foam systems suitable for thermal insulation of cavity walls with masonry or concrete inner and outer leaves* (16 pages). This British Standard specifies the property requirements, the properties of the components and the production parameters, of urea-formaldehyde foam systems suitable for injection into external masonry or concrete cavity walls to provide improved thermal insulation. Owing to the problems that have been encountered with this product in the past (incorrect production and application can lead to the release of toxic fumes into the property) it is seldom used today as a cavity insulant. As a result of this, it should also be installed in accordance with BS5618: 1985 *Code of practice for thermal insulation of cavity walls (with masonry or concrete inner and outer leaves) by filling with urea-formaldehyde (UF) foam systems* (62 pages). This British Standard code of practice describes recommendations for the installation of UF foam systems which are dispensed on site, to fill the cavities of suitably situated and constructed external walls of maximum height 12 m, which have masonry or concrete inner and outer leaves, thereby providing additional thermal insulation to such walls. It defines what are suitably situated and constructed external walls and indicates essential procedures and precautions for the filling process (e.g. before work is commenced the wall should be assessed for suitability and the work should be carried out by a person operating under an Approved Installer Scheme that includes an assessment of capability). Walls built of random rubble are not covered by this code. Because of the health risks associated with the use of UF foam, it is subject to its own Part of the Building Regulations (see *Building Regulations 2000*, Part D and Approved Document D).
- Other insulating materials inserted after the wall has been constructed should have certification from an appropriate body and be installed in accordance with

the appropriate installations code. Before work is commenced the wall should be assessed for suitability and the work should be carried out by a person operating under an Approved Installer Scheme that includes an assessment of capability. Alternatively, the insulating material should be the subject of current certification from an appropriate body or a European Technical Approval. The work should be carried out to meet the requirements of the relevant document by operatives directly employed by the document holder. Alternatively, they may be employed by an installer approved to operate under the document.

- Where materials are being inserted into a cavity wall of an existing house, the suitability of the wall for filling should be assessed, before installation, in accordance with BS 8208: Part 1: 1985 *Guide to assessment of suitability of external cavity walls for filling with thermal insulation* (30 pages). This Part of BS 8208 gives guidance on factors to be considered when assessing the suitability of existing external cavity walls with masonry and/or concrete leaves for filling with thermal insulants. It applies to cavity walls where:
  - the height of the cavity wall does not exceed 12 m (see Clause 2.4)
  - vertical members of structural frames do not bridge the cavity (see Clause 2.1)
  - there is no existing cavity insulation.

  Special attention should be given to the condition of the external leaf of the wall, (e.g. its state of repair and type of pointing). Some materials that are used to fill existing cavity walls may have a low risk of moisture being carried over to the internal leaf of the wall. In cases where a third party assessment of such a cavity fill material contains a method of assessing the construction of the walls and exposure risk, the procedure set out in Diagram 12 (see Figure 4.5) and Table 4 (see Table 7.1) may be replaced by that method. This standard is intended for people who are appropriately qualified and experienced and are properly trained to carry out assessments of cavity walls for filling with insulants. Additionally, because certain features of a building may only become apparent at the time of carrying out the work, it is essential also that installation technicians are trained in assessment.

Further information regarding the insulation of external walls is illustrated in Figure 7.4.

7.4.4 Impervious claddings for external walls

The principles of external claddings are illustrated in Figure 7.2(b) and (c) where it is shown that they should:

- resist the passage of precipitation to the inside of the building
- not be damaged by precipitation

- not transmit moisture due to precipitation to any part of the building that might be damaged.

It is possible to design cladding so that it can either hold precipitation (even if wind driven) at the face of the building (impervious cladding), or allow precipitation to penetrate beyond the face but stop it from getting beyond the back of the cladding.

Therefore the cladding should be either:

(a) jointless (or have sealed joints) and be impervious to moisture (such as sheets of metal, glass, plastic or bituminous materials), or

(b) have overlapping dry joints and consist of impervious or weather-resisting materials (such as natural stone or slate, cement based products, fired clay or wood).

Dry jointed claddings should be backed by a material (such as sarking felt) which will direct any penetrating moisture to the outside surface of the structure.

Moisture-resisting materials consisting of bituminous or plastic products lapped at the joints are permitted but they should be permeable to water vapour unless there is a ventilated space behind the cladding.

Materials that are jointless or have sealed joints should be designed to accommodate structural and thermal movement.

Dry joints between cladding units should be designed either to resist precipitation or to direct any moisture entering them to the outside face of the structure. The suitability of dry joints will depend on the design of the joint and cladding and the severity of exposure of the building to wind and rain.

All external claddings should be securely fixed and particular care should be taken with detailing and workmanship at the junctions between window and door openings and the cladding since these junctions are particularly susceptible to the ingress of moisture. BS 8000: Part 6: 1990 *Workmanship on building sites. Code of practice for slating and tiling of roofs and claddings* (30 pages) may be of some assistance for claddings containing these materials. Part 6 of BS 8000 gives recommendations on basic workmanship and covers those tasks which are frequently carried out in relation to slating and tiling of roofs and claddings of buildings.

The recommendations apply to the laying and fixing of clay and concrete tiles, natural and fibre-reinforced cement slates and their associated fittings and accessories. BS 8000 as a whole is unique in that, unlike other British Standards, it draws together recommendations given in other codes of practice. Its purpose is to encourage good workmanship by providing the following:

- the most frequently required recommendations on workmanship for building work in a readily available and convenient form to those working on site

- assistance in the efficient preparation and administration of contracts
- recommendations on how designer's requirements for workmanship may be satisfactorily realized
- definitions of good practice on building sites for supervision and for training purposes (Note: this guidance is not intended to supplant the normal training in craft skills)
- a reference for quality of workmanship on building sites.

At present BS 8000 comprises the following Parts:

Part 1 Code of practice for excavation and filling
Part 2 Code of practice for concrete work
Part 3 Code of practice for masonry
Part 4 Code of practice for waterproofing
Part 5 Code of practice for carpentry, joinery and general fixings
Part 6 Code of practice for slating and tiling of roofs and claddings
Part 7 Code of practice for glazing
Part 8 Code of practice for plasterboard partitions and dry linings
Part 9 Code of practice for cement/sand floor screeds and concrete floor toppings
Part 10 Code of practice for plastering and rendering
Part 11 Code of practice for wall and floor tiling
Part 12 Code of practice for decorative wall-coverings and painting
Part 13 Code of practice for above ground drainage and sanitary appliances
Part 14 Code of practice for below ground drainage
Part 15 Code of practice for hot and cold water services (domestic scale).

Some materials, such as timber claddings, are subject to rapid deterioration unless properly treated. These materials should only be used as the weather-resisting part of a roof or wall if they can meet the conditions specified in Approved Document to support Regulation 7. As the name suggests, this approved document supports Regulation 7 of the *Building Regulations 2000* which states that:

Building work shall be carried out –

(a) with adequate and proper materials which –
   (i) are appropriate for the circumstances in which they are used,
   (ii) are adequately mixed and prepared, and
   (iii) are applied, used or fixed so as adequately to perform the functions for which they are designed; and
(b) in a workmanlike manner.

The Approved Document gives guidance on the choice and use of materials, and on ways of establishing the adequacy of workmanship. It should be noted, however, that under the building regulations, materials and workmanship are controlled only to the extent of:

- securing reasonable standards of health and safety for persons in or about buildings for Parts A to D, F to K, N and P of Schedule 1 (except for paragraphs H2 and J6)
- conserving fuel and power in Part L
- providing access to buildings for people in Part M.

It should be noted that the weather-resisting part of a roof or wall does not include paint or any surface rendering or coating which does not of itself provide all the weather resistance. Where the cladding is on the façade of a timber framed building or is itself supported by timber components the construction should be ventilated so that rapid drying out of any penetrating moisture is ensured. Furthermore, the cladding to framed external walls (see Figure 7.4(e) and (f)) should be separated from the insulation or sheathing by a vented and drained cavity with a membrane that allows water vapour to pass through it but resists the passage of liquid water on the inside of the cavity.

Insulation may be incorporated into the roof or wall cladding provided that it is protected from moisture (or is unaffected by it). Possible problems may arise due to interstitial condensation and cold bridges in the construction. Further guidance on this may be found in BRE Report BR 262 *Thermal insulation – avoiding risks* (see section 6.3.4 above).

7.4.5 *Impervious claddings for external walls – alternative approach*

Relevant recommendations of the following alternative guidance sources can be used instead of the somewhat generalized guidance given in Approved Document C:

- British Standard Code of Practice 143 *Code of practice sheet roof and wall coverings*. The BS Code of Practice 143 consists of a series of parts dealing with different sheet roofing and walling materials. The parts of the code referred to in Approved Document C are as follows:
  - Part 1: 1958 *Corrugated and troughed aluminium* (16 pages) – deals with aluminium alloy corrugated and troughed sheeting conforming to BS 2855

used as covering for roofs and sides of buildings. It contains information on weathering, contact with other materials, fire hazard and other characteristics, and makes recommendations relating to materials, design and construction.

○ Part 5: 1964 *Zinc* (22 pages) – deals with zinc roof covering and gives recommendations based on accepted good practice in the United Kingdom for laying zinc roofs on the Roll Cap System. In this system the sheets are fully supported and have their sides turned up against shaped wood rolls spaced to suit the width of the sheets. The rolls are covered with an independent capping which completes the joint. Joints between the ends of sheets are made either by means of drips or welts depending on the pitch. Recommendations are given for the whole of the roof covering above the upper surface of the constructional base for both flat and pitched roofs. Flashings and gutters are also dealt with in so far as they are integral parts of the roof covering.

○ Part 10: 1973 *Galvanised corrugated steel* (22 pages) – deals with the use of galvanized corrugated steel sheets for roofing and cladding in building. Recommendations are given on materials and design, construction and maintenance, together with information on weathertightness, durability, thermal insulation, fire hazard, rainwater drainage from roofs and other characteristics.

○ Part 12: 1970 (1988) *Copper* (30 pages) – deals with methods of covering a roof or wall with copper sheet or strip in accordance with established practice. There are alternative methods of laying which are not included in the techniques described, as they are generally variations of traditional roof practice. Recommendations are given in regard to the whole of the coverings above the wooden rafters for pitched roofs, and above the joists or the upper surface of the constructional base for flat roofs and walls. Flashings and gutters are dealt with in so far as they are integral parts of the covering.

○ Part 15: 1973 (1986) *Aluminium* (34 pages) – gives recommendations for the installation of aluminium fully supported roof and wall coverings in accordance with established practices. It includes information on appropriate alloys and forms of aluminium, durability, contact with other materials, sizes and weights of sheet and strip, protection and storage. Gutters and flashings are dealt with in so far as they form an integral part of the main roof covering. The code deals with coverings, substructures and accessories placed above the rafters or the upper surface of constructional bases, but does not apply to aluminium roof decking or deck units.

○ Part 16: 1974 *Semi-rigid asbestos bitumen sheets* (32 pages) – deals with semi-rigid asbestos bitumen sheets (referred to as SRABS) and gives recommendations based on accepted good practice for laying fully supported semi-rigid asbestos bitumen sheet roofs using the roll cap and rib systems. Recommendations are given for the whole of the roof covering above the upper surface of the constructional base for both flat and pitched roofs. Flashings and gutters are dealt with in so far as they are integral parts of the roof covering.

Although they are rather out of date now, the various parts of the code are easy to read and contain numerous drawings and details to illustrate the design and construction principles. They are of use to anyone concerned with the design, construction, repair and maintenance of sheet roof and wall coverings in traditional materials.

- BS 6915: 2001 *Specification for design and construction of fully supported lead sheet roof and wall coverings* (48 pages). This Standard gives recommendations for the design and construction of fully supported coverings of rolled lead sheet, conforming to BS EN 12588, applied to external wall and roof surfaces. The recommendations apply primarily to fixing the lead *in situ* and cover the design and fixing of the whole of the covering above the substructure that supports it, including the design and fixing of gutters and flashings. References are made to coverings that are based on the use of preformed panels to which the main principles of design also apply. General information on the properties of lead roof coverings is given in Annex A in the Standard. The recommendations of this Standard apply mainly to rolled lead sheet. Sand cast lead sheet, manufactured by established and reputable manufacturers, is produced in thicknesses and weights similar to codes 6, 7, 8 and 9 of BS EN 12588:1999 and is used mainly for replacement work on old and historic buildings and for ornamental leadwork. It has a relatively smooth surface one side and a sand-faced texture on the other. For all practical purposes where the composition of the metal is similar, there is no significant difference in the working or fixing recommendations of sand cast lead sheet and rolled lead sheet. Where the composition of the metal is different, the manufacturer should be consulted. It should be noted that the colour coding system used for rolled lead sheet does not apply to sand cast or direct method lead. This is a very useful document containing a large number of practical details and much useful advice. It will be of use to anyone concerned with the design, construction, specification, repair and maintenance of lead sheet roof and wall coverings.

- BS 8219: 2001 *Profiled fibre cement. Code of practice* (20 pages). This BS code of practice gives recommendations for design specific to the use of profiled fibre cement sheets for roof and wall cladding on buildings, and recommendations for basic workmanship and tasks carried out in relation to the installation of profiled fibre cement sheets for roofing and external wall cladding including associated fittings and accessories. The code takes into account the change to non asbestos reinforced fibre cement profiled sheets and accessories together with the introduction of proprietary self-drilling fastener systems. The code covers:

○ classification – fibre reinforced profiled sheets should be classified in accordance with BS EN 494 (the characteristics of profile shapes and longitudinal and cross breaking strengths are provided in Table 1 of the code)
○ checking, handling and working, site storage of materials and preparation
○ installation of sheets
○ safety and general precautions.

The code will be of use to anyone concerned with the design, construction, repair and maintenance of fibre cement sheet roof and wall coverings.

- BS 8200: 1985 *Code of practice for design of non-loadbearing external vertical enclosures of buildings* (84 pages). The purpose of this code of practice is to provide a systematic framework within which an enclosure can be designed and constructed. It is intended to be used by designers as a comprehensive checklist, incorporating useful guidance, and not for inclusion in specifications or other contract documents. In order to achieve this, the code makes extensive reference to other publications which cover detailed considerations of performance and design. The designer, whether he is designing a walling system for a specific building or group of buildings, or a component for general use, will need to refer to all sections of the code, using the process referred to in Section 3. In giving recommendations for the design of non-loadbearing external vertical enclosures of buildings, the code covers the following four main areas:

○ performance criteria – recommendations for establishing the performance required of the walling, using the data assembled in Appendices B to F (Appendix B Climate, Appendix C Locality, Appendix D Activity criteria, Appendix E Noise criteria, Appendix F Side effects of activities)
○ design – guidance on the design of the enclosure to satisfy the requirements established in Section 2 (Performance criteria). Section 3 (Design) when read with Appendix H, also contains a recommended process for the designer.
○ Workmanship and contract management: recommendations for ensuring that the design is realized on site.
○ Maintenance: recommendations for ensuring that the performance of the enclosure is maintained during its life.

The code is intended to be used primarily by designers to assist in the design process. Additionally it will assist the designer in his role of supervisor, assessor and adviser to his or her client. It will also be of use to those to whom a designer looks for support in the execution of the design, i.e. manufacturers, erector and main contractor.

- BS 8297: 2000 *Code of practice for design and installation of non-loadbearing precast concrete cladding* (48 pages). This BS code of practice gives recommendations and guidance for the design, manufacture, transport and installation of non-load-bearing precast concrete cladding in the form of:

o units supported by and fixed to a structural frame or wall
o units used as permanent formwork in part or in whole.

Recommendations are included on the precautions which have to be taken to enable the cladding to perform its function satisfactorily, such as the need to provide for permanent and temporary movements of the structure due to shrinkage and elastic deformation under load. The minimum standards needed are given as well as details of the materials and methods of fixings most frequently used. Although the code applies primarily to new buildings, many provisions may be applicable to alterations or refurbishment of existing buildings. Guidance is given on the manufacture of units and their surface finish and methods of test. It should be noted that the procedure for structural design given in the code is based on limit state design and an abbreviated method of determination of design wind loadings for low-rise buildings is given in Annex A of the code. The code covers:

o materials and components
o design of cladding units
o position and detail of joints
o support and attachment of units to the structure
o surface finish of cladding units
o manufacture
o handling and transportation of cladding units
o on-site erection and fixing
o performance testing of cladding units
o inspection and maintenance.

It does not cover load-bearing cladding, units incorporating glass reinforced cement, or the design of the supporting structure to which the cladding is attached. The code contains a great deal of practical information and will be of use to designers, contractors, manufacturers and regulators concerned with the use of non-load-bearing precast concrete panel construction.

• BS 8298: 1994 *Code of practice for design and installation of natural stone cladding and lining* (48 pages). This BS code of practice contains recommendations for the design, installation and maintenance of mechanically fixed facing units of natural stone as a cladding and lining where these are:

o held to a structural background by metal fixings, or
o attached to precast concrete units (i.e. stone faced concrete cladding units).

It does not include load-bearing cladding or cladding held only by adhesion, any type of cladding supported or held in position around the perimeter of stones or series of stones by metal framing, or the use of stone cladding as permanent formwork to *in situ* concrete. The code is divided into six sections as follows:

Section 1: General
Section 2: Materials and components
Section 3: Design
Section 4: Workmanship in production
Section 5: Workmanship on site
Section 6: Maintenance.

Recommendations concerning traditional stone masonry including ashlar may be found in BS 5390, although this document is not referred to in Approved Document C. This practical and informative code will be of use to all concerned with the design, installation and maintenance of mechanically fixed facing units of natural stone.

- MCRMA (Metal Cladding and Roofing Manufacturers Association) Technical Paper 6 *Profiled metal roofing design guide*, revised edition 2004 (32 pages). Profiled steel or aluminium sheets are used in various roof constructions such as single-skin and double-skin systems, secret fix, site-assembled or factory-made composite panels and under-purlin linings. The guidance in the technical paper applies in principle to all of these constructions; however, it is particularly aimed at the double-skin system with trapezoidal or secret fix profiled sheets (the most common type of metal roof construction used in the UK). Sinusoidal profiles (e.g. corrugated iron) are not used for modern industrial and commercial roofs and they are not covered by the guide. Secret fix systems and composite panels (see next reference) are both described fully in separate MCRMA technical guides. The technical paper covers the following:

  1.0 Typical construction and assemblies
  2.0 Components
  3.0 Weathertightness
  4.0 Thermal Performance
  5.0 Interstitial condensation
  6.0 Acoustics
  7.0 Performance in fire
  8.0 Structural performance
  9.0 Durability
  10.0 Sustainability
  11.0 Construction details and accessories
  12.0 Site work

13.0 Inspection and maintenance

14.0 References.

This excellent paper provides non-technical and highly readable information which will be of use to both specialists and non-specialists alike. It is well illustrated and covers basic principles of metal roof design. One note of caution – the section on thermal performance is now out of date and Approved Document C refers to the 1996 edition of this paper since it predated the latest revision.

- MCRMA Technical Paper 9 *Composite roof and wall cladding panel design guide*, 1995 (20 pages). This paper covers the materials, methods of manufacture and performance of metal composite roof and wall cladding panels. Whilst it may be of use in giving an indication of the general principles for the design and installation of these products the paper is now well out of date, especially with regard to thermal performance. The reader would be better advised consulting individual manufacturers where often excellent design details may be freely downloaded from the Internet.

7.5 Cracking of external walls

This very brief section which was added into the 2004 edition of Approved Document C is curious in that it rather states the obvious, that severe rain penetration may occur through cracks in masonry external walls! It goes on to give the reasons for the cracking as being thermal movement in hot weather or subsidence after prolonged droughts and advises that this should be taken into account when designing the building. There are of course many causes of cracking in external walls including moisture movement in clay brickwork, cavity wall tie failure, sulphate attack, frost action, swelling and shrinking of clay soils, mining subsidence, faulty drains, etc. etc. In order to assist the reader the following reference sources are given in the Approved Document:

- *BRE Building Elements. Walls, windows and doors* (*Performance, diagnosis, maintenance, repair and the avoidance of defects*) 1998 (302 pages). This excellent, well illustrated and informative book cannot be easily summarized in just a few paragraphs, however the preface gives an idea of the book's scope, as follows:
  - It covers all kinds of external walls, both loadbearing and non-load-bearing; windows and doors.
  - It describes both good and bad features of walls, windows and doors and the joints between them.

○ It concentrates on those aspects of construction which, in the experience of BRE, lead to the greatest number of problems or greatest potential expense, if carried out unsatisfactorily.

The book also contains a number of case studies in some of the chapters which have been selected from the files of the BRE Advisory Service and the former Housing Defects Prevention Unit. These represent the most frequent kinds of problems on which BRE has been consulted. The book is structured so that Chapter 1 contains a general discussion of principles whereas Chapters 2 to 10 contain practical information on a number of different external enclosure types. Cracking is dealt with throughout the book, as the causes and effects of this will depend on the type of walling being considered. The book is primarily aimed at building surveyors and other professionals performing similar functions (such as architects and builders) who maintain, repair, extend and renew the national building stock. It will also be of use to students.

- *BRE Report BR 292 Cracking in buildings*: 1996 (106 pages). The purpose of this report is to describe the science behind the different causes of cracking thereby enabling surveyors and others who investigate the phenomenon to understand the causes and hence be able to offer diagnoses and remedies. The report sets out basic information on the science of materials behaviour which is relevant to understanding how and why cracks occur. The report is structured as follows:
  ○ Part 1 – deals with the underlying science (i.e. the physics and chemistry) underlying the changes of size in materials and components. This also includes basic data (such as quantifying size changes and distortions in building materials) which is essential both in designing to avoid cracking damage and in the diagnosis of the causes of damage in existing structures. Part 1 goes on to describe the mechanisms by which the size changes potentially produce intolerable strain, and consequent distortion or cracks.
  ○ Part 2 – deals with the causes of cracking by setting them in real building contexts. Each building element is taken in turn and the information is presented in a common format, typically:
    – design principles
    – practical detailing
    – site practices
    – diagnostic principles
    – remedial work or repairs.

The report is aimed at three main interest groups:
  ○ construction professionals (architects, surveyors and contractors)
  ○ litigators (building failure investigators, loss adjusters and expert witnesses)
  ○ building owners and maintenance staff
to enhance the interest that they have in relation to their own particular role, and also to provide a general appreciation of the subject and some understanding

of the interests of the other parties in the building process. The report is highly readable and well illustrated and the way that it is set out means that it will be readily accessible to all members of the above groups.

- BS 5628: Part 3: 2001 *Code of practice for use of masonry. Materials and components, design and workmanship* (see sections 6.3.2, 7.4.1 and 7.4.2 above for more information on this Standard).

The first two references are both excellent information sources. The author has also found the publication *Common Defects in Buildings* published by The Stationery Office to be invaluable, especially in the diagnosis of defects with particular reference to cracking in external walls. BS 5628 is more relevant to choice of materials.

7.6 Detailing of joints between walls and door and window frames

In many cases, the most vulnerable part of the construction regarding moisture penetration due to precipitation is at the junction between the wall and any door, window or other openings into the building. The joint formed at such junctions should:

- resist the passage of precipitation to the inside of the building
- not transmit moisture due to precipitation to other parts of the building that might be damaged
- not be damaged by precipitation.

The usual way of satisfying these performance criteria is by the use of suitably placed damp-proof courses to direct moisture to the outside. These should be provided:

(a) where moisture flowing downward would be interrupted by an obstruction, such as a lintel
(b) where a sill under a door, window or other opening (including any associated joint) might not form a complete barrier to the transfer of precipitation
(c) where the construction of a reveal at a door, window or other opening (including any associated joint) might not form a complete barrier to the passage of rain or snow.

Modern methods of construction often lead to rather wide cavities due to the additional thicknesses of insulation now required by Part L (see Approved Document L

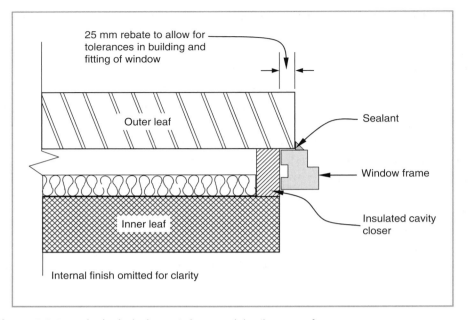

25 mm rebate to allow for
tolerances in building and
fitting of window

Outer leaf

Sealant

Window frame

Insulated cavity
closer

Inner leaf

Internal finish omitted for clarity

Figure 7.6 Example checked rebate window reveal detail – areas of severe or very severe exposure to driving rain

for guidance) and the need in some cases for a drained 50 mm wide cavity in addition to the insulation where the cavity is only partially filled (see section 7.4.3 above). In such cases most window and door frames will not fully cover the cavity closer. This will necessitate lining the reveal with plasterboard, dry lining, a thermal backing board or even a support system. The internal reveal may also be wet plastered provided that the plaster is applied to a suitable backing such as expanded metal lathing or similar support system.

In areas of the country where the exposure to driving rain is in zone 4 (see section 7.4.3 above and Figure 7.5) additional precautions should be taken to prevent the penetration of precipitation. Approved Document recommends the use of checked rebates at all window and door frames. In this form of construction (Figure 7.6) the frame is set back behind the outer leaf of masonry. An alternative to this might be by the use of a proprietary finned cavity closer.

7.7 Door thresholds

Approved Document M, *Access to and use of buildings* (see *Using the Building Regulations, Part M Access*, first edition 2006), specifies that in general the access into a dwelling or block of flats from the outside should be provided with an accessible threshold, irrespective of whether the approach is level, ramped or stepped. Exceptionally, if the approach has to be stepped and for practical reasons a step into the dwelling is unavoidable, the rise should be no more than 150 mm. Clearly, this has implications for the weatherproofing of the building since there is a danger that precipitation and groundwater could enter the building at door openings. In order to prevent this, Approved Document C recommends that the external landing should slope away from the doorway at a fall of between 1 in 40 and 1 in 60. In order to encourage run-off, the door sill should have a maximum slope of 15°. Some additional recommendations are illustrated in Figure 7.7. Additional advice may be obtained from the following publications:

- BRE Good Building Guide 47 *Level external thresholds: reducing moisture penetration and thermal bridging,* 2001 (8 pages). This Good Building Guide describes some of the technical risks associated with the design of level thresholds and some detailing solutions. The guide gives advice on designing level

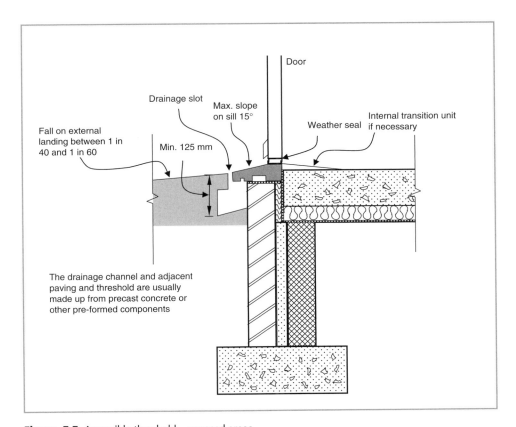

Figure 7.7 Accessible threshold – exposed areas

thresholds, not only for wheelchair users but also for those with walking difficulties, who would find even small steps and threshold bars extremely difficult. Minimum spatial dimensions are indicated (taken from Approved Document M) and the three principal elements of an accessible or level threshold are described and illustrated as:

○ the external landing and its drainage
○ the threshold, the sill and its junction with the external landing
○ the internal junction of the threshold with the floor finish.

The BRE Good Building Guides have been developed to provide practitioners with concise guidance on the principles and practicalities for achieving good quality building. The guides are designed to encourage and improve mutual awareness of the roles of different trades and professions. They are well illustrated and are written in clear non-technical terms and are suitable for anyone concerned with the built environment.

• *Accessible thresholds in new buildings: guidance for house builders and designers*. TSO, 1999 (23 pages). The aim of this guide is to:

○ suggest design solutions that will make the thresholds of dwellings more accessible to wheelchair users and people with limited mobility, whilst minimizing the risk of water entering the building
○ help designers achieve solutions that do not conflict with other aspects of Building Regulations.

The guide sets out the objectives in terms of the function or purpose of the main elements of the threshold and its surroundings, and then gives guidance on possible solutions. These solutions are not exhaustive because of the large number of combinations of elements and components involved. However, they are based on the experience of housing providers and builders, as well as individuals and groups concerned with disability issues. The guide is structured as follows:

Section 1: The external landing and its drainage – landing size, landing surface, landing drainage, drainage slots and channels
Section 2: The threshold, the sill and its junction with the external landing – drained landings, landings without drainage, timber sills, sill profile, weatherproofing, detailing beneath the sill
Section 3: The internal junction of the threshold with the floor finish – floor finish level, internal junctions
Appendix: Example threshold details.

This is the principal reference for the design of accessible thresholds and is a necessity for all who are concerned with the design, construction and regulation of dwellings.

7.8 Resistance to damage from interstitial condensation for external walls

In general, external walls will satisfy the requirements of the Building Regulations regarding resistance to damage from interstitial condensation if they are designed and constructed in accordance with:

- Clause 8.3 of BS 5250: 2002 *Code of practice for the control of condensation in buildings* (see section 6.3.4 above for more details of this code). Clause 8.3 describes ways of minimizing harmful interstitial condensation within a wall by:
  - specifying materials of decreasing vapour resistance from inside to outside where possible
  - avoiding service openings through a vapour control layer. Where this is not possible, they should be kept to a minimum and any openings taped and/or sealed
  - providing vented airspaces with openings to the outside
  - ensuring that external claddings which require the use of a membrane to avoid rainwater penetration (or a membrane to protect the insulation), are of the breather type meeting the requirements of BS 4016,

or

- BS EN ISO 13788:2001 *Hygrothermal performance of building components and building elements. Internal surface temperature to avoid critical surface humidity and interstitial condensation. Calculation methods* (see section 6.3.4 above for more details of this Standard).

In swimming pools and other buildings where high levels of moisture are generated, there is a particular risk of interstitial condensation in walls and roofs (see 8.4 below) because of the high internal temperatures and humidities that exist. In these cases specialist advice should be sought when these are being designed.

7.9 Resistance to surface condensation and mould growth in external walls

In order to resist surface condensation and mould growth in external walls it is necessary to ensure that the surface is maintained above the dewpoint temperature. This will depend on the outside temperature, the temperature of the room in which the wall surface is situated and the relative humidity of the room. It can be affected by ventilation, by thermal bridging of construction elements at junctions and by the degree of thermal insulation provided in the wall elements. Therefore, in external walls, care should be taken to design the junctions between the elements so that thermal bridging is avoided. This can be done by following the recommendations in the report *Bridging and air leakage: robust construction details for dwellings and similar buildings*, published by The Stationery Office (TSO, 2001) (see section 6.3.5 above) or by following the guidance of BRE Information Paper IP17/01 *Assessing the effects of thermal bridging at junctions and around openings* (see section 6.3.5 above). Additionally, externals walls should be designed and constructed so that the thermal transmittance (U-value) does not exceed 0.7 W/m^{2}K at any point.

Roofs

8.1 Regulation requirements

Requirement C2 of Schedule 1 to the *Building Regulations 2000* deals with resistance to moisture. It requires the floors, walls and roof of a building to adequately protect it and the people who use it from harmful effects caused by:

- moisture from the ground
- precipitation and wind-driven spray
- surface and interstitial condensation
- water spilt from or associated with sanitary fittings and fixed appliances.

Guidance on the application of Requirement C2 to walls is given in Approved Document C in Section 5.

8.2 Definitions

In Approved Document C the following definitions apply with regard to Requirement C2:

FLOOR – the lower horizontal surface of any space in a building including any surface finish which is laid as part of the permanent construction. This would, presumably, exclude carpets, linoleum, tiles, etc., but would include screeds and granolithic finishes.

GROUNDWATER – liquid water (i.e. not water vapour, ice or snow, etc.), either flowing through the ground or as a static water table.

INTERSTITIAL CONDENSATION – water vapour being deposited as liquid water within or between the layers of the envelope of the building.

MOISTURE – water present as a liquid, gas (e.g. water vapour) or solid (e.g. ice or snow, etc).

PRECIPITATION – moisture in any form falling from the atmosphere, such as rain, sleet, snow or hail, etc.

ROOF – any part of the external envelope of a building that makes an angle of less than 70° to the horizontal.

SPRAY – wind-driven droplets of water blown from the surface of the sea or other bodies of water close to buildings. The salt content of sea spray makes it especially hazardous to many building materials.

SURFACE CONDENSATION – water vapour being deposited as liquid water on visible surfaces within the building.

VAPOUR CONTROL LAYER – Typically, this is a membrane material which is used in the construction, with the purpose of substantially reducing the transfer of water vapour through any building in which it is incorporated.

WALL – any opaque part of the external envelope of a building that makes an angle of 70° or more to the horizontal.

8.3 Weather resistance of roofs

The roof of a building should:

- resist the penetration of precipitation to the inside of the building
- not be damaged by precipitation
- not transmit precipitation to another part of the building that might be damaged
- be designed and constructed so as not to allow interstitial condensation to adversely affect its structural and thermal performance
- not promote surface condensation and mould growth under reasonable occupancy conditions.

The recommendations for impervious external wall claddings mentioned in section 7.4.4 above apply equally to roof covering materials.

The performance requirements for external wall and roof claddings can also be met if they comply with:

- British Standard Code of Practice 143 *Code of practice sheet roof and wall coverings* (this includes recommendations for aluminium, zinc, galvanized corrugated steel, copper and semi-rigid asbestos bitumen sheet)
- BS 6915: 2001 *Specification for design and construction of fully supported lead sheet roof and wall covering*
- BS 8219 *Profiled fibre cement. Code of practice*

- BS 8200: 1985 *Code of practice for design of non-loadbearing external vertical enclosures of buildings*
- MCRMA Technical Paper 6 *Profiled metal roofing design guide*, revised edition 1996
- MCRMA Technical Paper 9 *Composite roof and wall cladding panel design guide*, 1995.

The above documents (which are described in section 7.4.5) contain details of the materials to be used and contain design guidance including fixing recommendations.

8.4 Resistance to damage from interstitial condensation in roofs

This section was originally contained in Part F *Ventilation* of Schedule 1 to the *Building Regulations 2000* – requirement F2. With the extension of Part C to cover condensation risk in the 2004 amendment it was logical to transfer these provisions to Part C and they have been omitted from the 2006 edition of Part F.

The guidance given in the current Approved Document C is much less detailed than that originally contained in Approved Document F and is now stated in performance terms and by reference to other sources of guidance where particular design details are illustrated. The current trend of continually referring to additional sources of guidance is making the Approved Documents much less useful to the practitioner as, in many cases, they can only be regarded as a directory and not as a direct source of technical guidance in their own right.

When condensation occurs in roof spaces it can have two main effects:

- the thermal performance of the insulant materials may be reduced by the presence of the water
- the structural performance of the roof may be affected due to increased risk of fungal attack.

Approved Document C recommends that interstitial condensation in roofs should be limited such that the thermal and structural performance of the roof will not be adversely affected.

It should be noted that the provisions of AD C apply to roofs of any pitch even though a roof which exceeds 70° in pitch is required to be insulated as if it were a wall. Additionally, small roofs over porches or bay windows, etc., may sometimes be excluded from the requirements of requirement C2(c) if there is no risk to health or safety.

In swimming pools and other buildings where high levels of moisture are generated, there is a particular risk of interstitial condensation in walls and roofs because of the high internal temperatures and humidities that exist. In these cases specialist advice should be sought when these are being designed.

For cold deck roofs (where the insulation is placed at ceiling level and can be permeated by moisture from the building) requirement C2(c) can be met by the provision of adequate ventilation in the roof space. In such roofs it is obviously essential that moist air is prevented from reaching the roofspace where it might condense within or on the insulation layer and this is particularly important above areas of high humidity such as bathrooms and kitchens. Weak points in the construction where this might occur include:

- gaps and penetrations for pipes and electrical wiring (these should be filled and sealed)
- loft access hatches (where an effective draught seal should be provided to reduce the inflow of warm moist air).

The requirements can be met for both flat and pitched roofs by following the relevant recommendations of:

- Clause 8.4 of BS 5250: 2002 *Code of practice: the control of condensation in buildings*. This code is described in general in section 6.3.4 above. Clause 8.4 recommends that the following should be considered in order to avoid interstitial condensation:
  - extract moisture at source (to reduce the risk of water vapour transferring from occupied areas to the roof)
  - minimize water vapour penetration to the cold side of a roof construction by keeping constructional gaps and holes in ceilings to a minimum and by sealing holes for service openings after the service has been installed
  - make sure that vapour control layers are adequately lapped and sealed and their integrity is maintained, and seal around the perimeter of the vapour control layers and at junctions and penetrations caused by such things as light cables, service runs and access hatches
  - place ventilation openings on the longer sides of a typical rectangular roof to achieve adequate cross ventilation
  - ensure that where ventilation openings are sited at intervals, they are of equivalent area to the continuous openings recommended and avoid stagnant air pockets due to inadequate through-flows and where possible exceed the minimum opening areas given in the codes
  - design ventilation openings so that they:
    - cannot be blocked (e.g. by dust, airborne debris, paint or frost)
    - prevent ingress of rain, snow, birds and large insects.

The code recommends that a nominal mesh/grill size of 4 mm be adopted (this also avoids excessive airflow resistance)

o consider using proprietary ventilators where these will avoid problems, or be more practicable in use

o maintain minimum defined free airspaces throughout both ventilation products and roof voids

o take care to prevent moisture from wet processes used in the roof construction (or rain during construction) from being trapped between the waterproof roof covering and a vapour control layer positioned at a lower level.

- BS EN ISO 13788: 2001 *Hygrothermal performance of building components and building elements. Internal surface temperature to avoid critical surface humidity and interstitial condensation. Calculation methods* (see section 6.3.4 above for more details of this Standard)

- Further guidance may also be found in the 2002 edition of BRE Report BR 262 *Thermal insulation: avoiding the risks.* This report is discussed in general in section 6.3.4 above. Advice on avoiding condensation in roofs is given in Section 2 of the report under the following headings:

2.1 Condensation within roof spaces

2.2 Condensation on the inside of extract ducts passing through unheated roof spaces

2.3 Condensation at thermal bridges

2.4 Risks associated with electrics

2.5 Freezing of water in pipes and cisterns

2.6 Increased heat loss where insulation is not full depth.

Much of this guidance is reproduced from BS 5250 and is discussed below.

The following guidance is based on that which was originally contained in Approved Document F which was in turn based on the guidance provided in BS 5250. It covers only a few typical examples of commonly found forms of construction where the insulation is placed at ceiling level (cold deck roofs). For full details and for other forms of construction reference should be made to BS 5250 Clause 8.4 (see above).

8.4.1 Roofs with a pitch of 15° or more

Pitched roofs should be cross-ventilated by permanent vents at eaves level on the two opposite sides of the roof, the vent areas being equivalent in area to a continuous gap along each side of 10 mm width. Ridge ventilation equivalent to

a continuous gap of 5 mm should also be provided where the pitch exceeds 35°
and/or the spans are in excess of 10 m.

Mono-pitch or lean-to roofs should have ventilation at eaves level as above,
and at high level either at the point of junction or through the roof covering at the
highest practicable point. The high-level ventilation should be equivalent in area
to a continuous gap 5 mm wide (see Figure 8.1).

In recent years vapour permeable underlays have come onto the market. If these
are used without continuous boarding below the tiling battens it is not necessary to
ventilate the roof space below the underlay. However, BS 5250 recommends that
in these circumstances, a ventilated space should be formed above the underlay
by using 25 mm battens and counterbattens and provision of ventilation at low
level equivalent to a 25 mm continuous gap and high-level equivalent to a 5 mm
continuous gap. Simply relying on fortuitous ventilation through the tile/slate joints
should not be relied on to ventilate this space adequately.

8.4.2 Roofs with a pitch of less than 15°

In low-pitched roofs the volume of air contained in the void is less and therefore
the risk of saturation is greater.

This also applies to roofs with pitch greater than 15° where the ceiling follows
the pitch of the roof. The high level ventilation should be equivalent in area to a
continuous gap 5 mm wide.

Cross-ventilation should again be provided at eaves level but the ventilation gap
should be increased to 25 mm width.

8.5 Resistance to surface condensation and mould growth in roofs

In order to resist surface condensation and mould growth in roofs and roof spaces,
care should be taken to design the junctions between the elements and the details
of openings, such as windows, so that thermal bridging and air leakage is avoided.
This can be done by following:

• the recommendations in the report *Limiting thermal bridging and air leakage:
  robust construction details for dwellings and similar buildings*, published by
  The Stationery Office (TSO, 2001) (see above, section 6.3.5). The Robust
  Details cover constructions of the following general types (the parts that cover

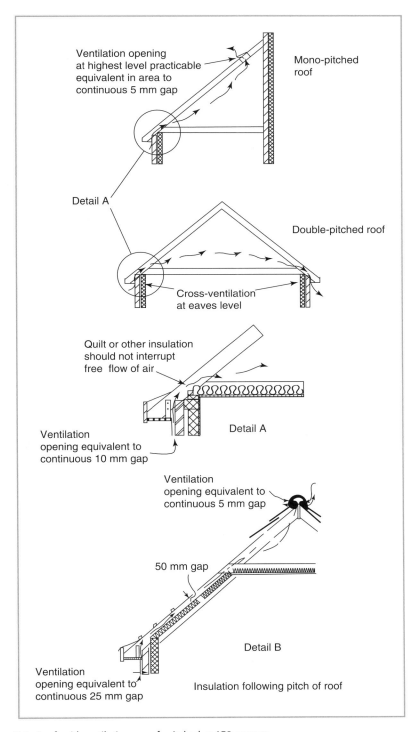

Figure 8.1 Roof void ventilation – roofs pitched at 15° or more

roof details are in brackets):

Section 2. Masonry: External Wall Insulation (parts 2.01 to 2.08)
Section 3. Masonry: Cavity Wall Insulation: Full-Fill (parts 3.01 to 3.08)
Section 4. Masonry: Cavity Wall Insulation: Partial-Fill (parts 4.01 to 4.08)

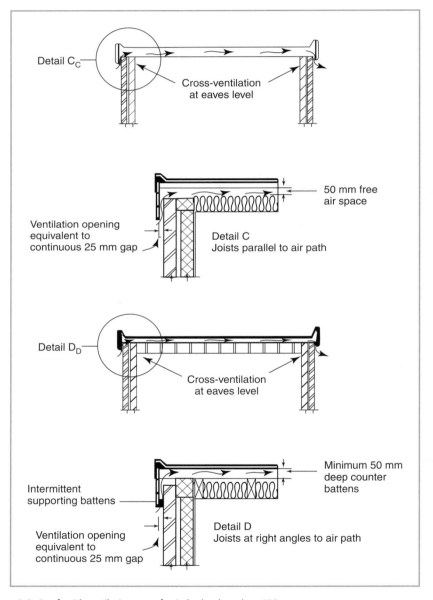

Figure 8.2 Roof void ventilation – roofs pitched at less than 15°

Section 5. Masonry: Internal Wall Insulation (parts 5.01 to 5.08)
Section 6. Timber Frame (parts 6.01 to 6.08)
Section 7. Light Steel Frame (parts 7.01 to 7.04),

or

- the guidance of BRE Information Paper IP17/01 *Assessing the effects of thermal bridging at junctions and around openings* (see section 6.3.5 above),

and/or

- the guidance in MCRMA Technical Paper 14 *Guidance for the design of metal roofing and cladding to comply with approved document L2: 2001*, 2002 revision (44 pages). This publication produced by the Metal Cladding and Roofing Manufacturers Association (MCRMA) in collaboration with BRE, was originally intended to assist designers, manufacturers and installers of metal walls and roofs to comply with the requirements of Part L of the Building Regulations which came into force in April 2002. Since those regulations have now been superceded by the 2006 revision of Part L and the 2006 edition of Approved Document L (see L2A and L2B) those parts of the 2001 edition quoted in Approved Document C are now obsolete. Until a new edition is published there is little point in obtaining the 2001 edition although the sections in the technical paper on infrared surveys and air leakage testing are of some general interest. The guidance is designed to be relevant to both twin skin and composite panel systems.

Additionally, roofs should be designed and constructed so that the thermal transmittance (U-value) does not exceed 0.35 W/m^2K at any point.

A free airspace of at least 50 mm should be provided between the roof deck and the insulation. This may need to be formed using counter-battens if the joists run at right angles to the flow of air (see Figure 8.2).

Where it is not possible to provide proper cross-ventilation an alternative form of roof construction should be considered.

It is possible to install vapour checks (called vapour control layers in BS 5250) at ceiling level using polythene or foil-backed plasterboard, etc., to reduce the amount of moisture reaching the roof void. This is not acceptable as an alternative to ventilation unless a complete vapour barrier is installed.

PART 4

Appendix

Appendix: Summary of references

Note: Many references are repeated throughout Approved Document C. In this summary we have tended to note the first appearance only of each reference.

Full title of reference	Reference location in Approved Document C	Reference location in this book	Target readership	Nature of guidance	Level of expertise required
Environmental Protection Act 1990 Part IIA (sections 78A–78YC)	Section 0 paragraph 0.9	Section 2.5	Regulators and professionals dealing with contaminated land issues.	Act of Parliament.	General appreciation of legal texts.
Environmental Protection (Duty of Care) Regulations 1991	Section 0 paragraph 0.9	Section 2.5	Regulators and professionals dealing with contaminated land issues. Companies moving contaminated waste.	Statutory Instrument.	General appreciation of legal texts.
Planning Policy Guidance Note PPG 23 *Planning and pollution control*, ODPM, 1997	Section 0 paragraph 0.10	Section 2.5	Regulators and professionals dealing with contaminated land issues.	Planning guidance document which gives guidance on the relevance of pollution controls to the exercise of planning functions.	General appreciation of legal guidance documents.

BS 7913: 1998 *Guide to the principles of conservation of historic buildings*	Introduction page 7	Section 2.6	Building owners, managers, archaeologists, architects, engineers, surveyors, contractors, conservators and local authority building control officers.	Designed to provide target audience with general background information on the principles of the conservation of historic buildings, when considering conservation policy, strategy and procedures.	Descriptive and informative – non-technical.
SPAB Information Sheet 4 – *The need for old buildings to breathe*, 1986	Introduction page 7	Section 2.6	All concerned with the repair, maintenance and upkeep of old buildings.	Aims to examine the differences between traditional and modern materials when used for maintenance purposes and to draw conclusions about the way old buildings should be treated.	Descriptive and informative – non-technical.

BRE Report BR 267 *Major alterations and conversions: a BRE guide to radon remedial measures in existing dwellings*, 1994	Introduction page 8	Section 2.6	All concerned with carrying out work to existing dwellings in radon-affected areas.	Offers advice on radon-protective measures that can be taken during the planning and implementation of major alteration or conversion works to buildings in radon-affected areas.	Descriptive and informative – does not require high levels of technical expertise.
British Standard 5930:1999 *Code of practice for site investigations*	Section 1 paragraph 1.2	Section 3.1.1	Key document for practitioners in the field of site investigation. May also be of use to designers, building control staff and general builders in giving a general understanding of the scope of site investigation and the techniques of ground investigation.	Deals with the investigation of sites for the purposes of: • assessing their suitability for the construction of civil engineering and building works, and • gaining knowledge of the characteristics of a site that affect the design and	Sections 1 to 3 descriptive and informative not requiring high levels of technical knowledge. Subsequent sections require specialist soils engineering knowledge.

		Sections 1 and 2 will found most useful in augmenting the limited information given in Approved Document C and described in section 3.1.	construction of civil engineering and building works and the security of neighbouring land and property.
	Section 3.1.1	Particularly useful for the non-specialist (and for students).	Non-technical giving practical advice on various issues concerned with site investigation.
BRE Digest 322 *Site investigation for low-rise building: procurement,* 1987	Section 1 paragraph 1.2		Procurement of the site investigation work, e.g. the value of site investigation, the steps that should be involved, and the contractual methods that can be employed to engage suitable specialists to carry out the work.

BRE Digest 318 *Site investigation for low-rise building: desk studies*, 1987	Section 1 paragraph 1.2	Section 3.1.1	Particularly useful for the non-specialist (and for students).	Discusses the influence desk studies can have on the identification of ground problems and contains a very useful checklist covering questions which need to be answered in order to assess which ground problems might occur.	Non-technical giving practical advice on various issues concerned with site investigation.
BRE Digest 348 *Site investigation for low-rise building: the walk-over survey*, 1989	Section 1 paragraph 1.2	Section 3.1.1	Particularly useful for the non-specialist (and for students).	Describes objectives of the walk-over survey (to check and make additions to information already collected during desk study) and shows the importance of visiting the site and its surrounding area and covering it carefully on foot.	Non-technical giving practical advice on various issues concerned with site investigation.

BRE Digest 381 *Site investigation for low-rise building: trial pits, 1993*	Section 1 paragraph 1.2	Section 3.1.1	Particularly useful for the non-specialist (and for students).	Describes how shallow trial pits can provide an economic and versatile way of examining soil conditions in situ.	Non-technical giving practical advice on various issues concerned with site investigation.
BRE Digest 383 *Site investigation for low-rise building: soil description, 1993*	Section 1 paragraph 1.2	Section 3.1.1	Particularly useful for the non-specialist (and for students).	Explains how to make an accurate description of the soil.	Non-technical giving practical advice on various issues concerned with site investigation.
BRE Digest 411 *Site investigation for low-rise building: direct investigations, 1995*	Section 1 paragraph 1.2	Section 3.1.1	Particularly useful for the non-specialist (and for students).	Repeats, updates and consolidates a great deal of the information provided by Digests described above and is most useful if a general idea of the scope and nature of site investigation is required.	Non-technical giving practical advice on various issues concerned with site investigation.

BS 8103: Part 1:1995 *Structural design for low rise buildings*	Section 1 paragraph 1.2	Section 3.1.1	May be of use to designers, building control staff and general builders.	Gives recommendations for the structural design of low-rise housing and covers the stability of the structure, site investigation and foundations and ground floor slabs used in the construction.	Descriptive and informative but requires a certain amount of technical expertise and an understanding of low-rise building construction.
BRE Digest 298 *Low-rise building foundations: the influence of trees in clay soils*, 1999	Section 1 paragraph 1.5	Section 3.2.2	Highly recommended for anyone concerned with the design and construction of low-rise buildings on shrinkable clay soils.	Gives simple guidance on minimising these effects in clay soils and discusses some dangers in current foundation practice.	Descriptive, informative and non-technical.

BRE Digest 241 *Low rise buildings on shrinkable clay soils: Part 2*, 1993	Section 1 paragraph 1.5	Section 3.2.2	Highly recommended for anyone concerned with the design and construction of low-rise buildings on shrinkable clay soils.	Reviews the use of traditional and trench fill strip foundations on sites containing shrinkable clay soils where trees and other vegetation are present.	Descriptive and informative. Does not require high levels of technical expertise.
Subsidence damage to domestic buildings: lessons learned and questions remaining, Foundation for the Built Environment (FBE), 2000	Section 1 paragraph 1.5	Section 3.2.2	Mainly of use to building surveyors and companies specialising in repair of subsidence damage.	Contains some useful research evidence on the nature of the problem of subsidence, and reviews the findings and contents of some of the Digests mentioned above. Lacking in technical detail.	Descriptive and informative. Non-technical background reading.

National House Building Council (NHBC) Standards Chapter 4.2 *Building near trees*, 2003	Section 1 paragraph 1.5	Section 3.2.2	Essential reading for designers, building contractors and regulatory bodies.	Practical guidance on assessing the risks of building near trees.	Descriptive and informative not requiring high levels of technical knowledge.
BRE Digest 427 *Low-rise buildings on fill* (Parts 1 to 3)	Section 1 paragraph 1.8	Section 3.2.3	Designers and building contractors	Gives guidance only of a general nature and it must be recognised that the services of civil, structural and geotechnical engineers will usually be needed.	Descriptive and informative but requires an understanding of the behaviour of fill materials.

BRE Report BR 424 *Building fill: Geotechnical aspects*	Section 1 paragraph 1.8	Section 3.2.3	Essential reading for all professionals who are specifically concerned with the appraisal of fill materials on site.	Provides a detailed account of BRE research findings and their significance for appropriate and successful building on fill.	Contains a great deal of highly technical ground engineering material which will only be of interest to trained engineers.
Defra/ Environment Agency Contaminated Land Research Report CLR 8 Potential contaminants for the assessment of land, 2002	Section 2 paragraphs 2.2 and 2.9, and Annex A	Section 4.1.1	Regulators, developers and their specialist advisers.	Identifies priority contaminants (or families of contaminants) and indicates which contaminants are likely to be associated with particular industries. Provides the DEFRA with a guide to the substances it should cover in its research work on contaminated land.	Mainly descriptive and informative. Useful as an introduction and guide to the other reports in this series (see below).

CLR 7 *Assessment of Risks to Human Health from Land Contamination: An Overview of the Development of Soil Guideline Values and Related Research* (DEFRA and Environment Agency, 2002)	Section 2 paragraph 2.9 and Annex A	Section 4.1.1	Regulators, developers and their specialist advisers.	Sets out the legal framework and the development and use of Soil Guideline Values, and has references to related research. Contains summaries of other reports in the series so useful background reading.	Mainly descriptive and informative.
CLR 9 *Contaminants in Soil: Collation of Toxicological Data and Intake Values for Humans* (DEFRA and Environment Agency, 2002)	Section 2 paragraph 2.9 and Annex A	Section 4.1.1	Regulators, developers and their specialist advisers.	Sets out the approach to the selection of tolerable daily intakes and Index Doses for contaminants to support the derivation of Soil Guideline Values.	Contains information of a specialist nature. Not for general reading.

CLR TOX 1–10 (DEFRA and Environment Agency, 2002)	Section 2 paragraph 2.9 and Annex A	Section 4.1.1	Regulators, developers and their specialist advisers.	These reports contain toxicological data used to derive Soil Guideline Values.	Contains information of a specialist nature. Not for general reading.
CLR 10 *The Contaminated Land Exposure Assessment Model (CLEA): Technical Basis and Algorithms* (DEFRA and the Environment Agency, 2002)	Section 2 paragraph 2.9 and Annex A	Section 4.1.1	Regulators, developers and their specialist advisers.	Describes the conceptual exposure models for each standard land-use that are used to derive the Soil Guideline Values.	Contains information of a specialist nature. Not for general reading.
CLR GV 1–10 (DEFRA and Environment Agency, 2002)	Section 2 paragraph 2.9 and Annex A	Section 4.1.1	Regulators, developers and their advisers.	These reports set out the derivation of the Soil Guideline Values for a range of contaminants.	Contains information of a specialist nature. Not for general reading.

CLR 11 *Model Procedures for the Management of Contaminated Land* (DEFRA and the Environment Agency, 2004)	Section 2 paragraph 2.9 and Annex A	Section 4.1.1	Regulators, developers and their specialist advisers.	Incorporates existing good technical practice into a systematic process for identifying, making decisions about and taking appropriate action to deal with contamination.	Descriptive and informative but highly specialised. Not for general reading.
Environment Agency R & D Technical Report P291 *Information on land quality in England: Sources of information (including background contaminants)*	Section 2 paragraph 2.3	Section 4.1.2	Of interest only to policy makers and developers, when making decisions about the quality of land which it is intended to develop.	An overview of information on land quality in England	Descriptive and informative but highly specialised. Not for general reading.

Environment Agency R & D Technical Report *P292 Information on land quality in Wales: Sources of information (including background contaminants)*	Section 2 paragraph 2.3	Section 4.1.2	Of interest only to policy makers and developers, when making decisions about the quality of land which it is intended to develop.	An overview of information on land quality in Wales.	Descriptive and informative but highly specialised. Not for general reading.
BRE Special Digest SD1 *Concrete in aggressive ground*, 2003	Section 2 paragraph 2.5	Section 4.1.2	Of use to concrete designers, contractors, specifiers and producers on the specification of concrete to resist chemical attack. Also of use to ground specialists in the assessment of ground in respect of aggressiveness to concrete.	Contains guidance on investigation, concrete specification and design to mitigate the effects of sulphate attack.	Descriptive and informative but some parts require knowledge of concrete chemistry.

HSE Report HSG 66 *Protection of workers and the general public during the development of contaminated land*, 1991	Section 2 paragraph 2.14	Section 4.2.4	Health and safety staff for all contractors and others working on contaminated sites.	Addresses the hazards associated with contaminated land working and aspects which developers and contractors need to consider under COSHH Regulations including the precautions to be taken during the development of the site.	Descriptive, informative and non-technical.
CIRIA Report 132 *A guide to safe working practices for contaminated land*, 1993	Section 2 paragraph 2.14	Section 4.2.4	Intended for use by a wide readership ranging from the general engineer to contaminated land specialists.	Provides guidance on safe working practices for contaminated sites.	Descriptive and informative. More detailed than the previous reference.

BS 10175: 2001 *Investigation of potentially contaminated land. Code of practice*	Section 2 paragraph 2.9	Section 4.2.5	Intended for use by those with some understanding of the risk-based approach to sites and site investigations, therefore more suitable for the specialist investigator.	Provides guidance on, and recommendations for, the investigation of potentially contaminated land or land with naturally enhanced concentrations of potentially harmful materials, to determine or manage the ensuing risks.	Descriptive and informative but mainly aimed at the specialist.
National Groundwater & Contaminated Land Centre report NC/99/38/2 *Guide to good practice for the development of conceptual models and the*	Section 2 paragraph 2.9	Section 4.2.5	Aimed at hydrogeologists and environmental professionals both internal and external to the Environment Agency.	Provides guidance on a generic 'good practice' approach to contaminant fate and transport modelling from setting objectives to interpretation of results and validation.	Specialist highly technical publication. Not for general reading.

Document	Section reference	Aimed at	Description	Comments
selection and application of mathematical models of contaminant transport processes in the subsurface				
Environment Agency R & D Technical Report P5-065 *Technical aspects of site investigation*, 2000 (Volumes I and II)	Section 2 paragraph 2.9 Section 4.2.5	Principally aimed at Environment Agency staff involved in the management of site investigation projects. Can also be of use to local authority officers and consultants and contractors engaged in site investigation projects.	Provides technical guidance on the investigation of contaminated sites for use in a wide variety of contexts.	Volume I – descriptive and informative and basically non-technical. Volume II – Well illustrated examples of standard forms and procedures.

Environment Agency R & D Technical Report P5-066 *Secondary model procedure for the development of appropriate soil sampling strategies for land contamination*	Section 2 paragraph 2.9	Section 4.2.5	Specialist document aimed at: • designers of soil sampling strategies • those who need to assess health and environmental risks on sites, and • those who rely on the output of the first two groups.	Contains explanatory material that discusses the relationship between soil sampling and risk assessment, and contains procedures for developing appropriate soil sampling strategies.	Descriptive and Informative but of specialist nature not for general reading.
Construction Industry Research and Information Association (CIRIA) Special Publication	Section 2 paragraph 2.20	Section 4.3.3	Designed to meet the needs of a wide range of potential users, including project and development managers,	Explains that containment and control can be achieved either by constructing suitable in-ground barriers or by using hydraulic measures.	Some sections are descriptive and informative, but it does contain a large amount of specialist information.

SP124 *Barriers, liners and cover systems for containment and control of land contamination,* 1996		consultants and contractors acting on behalf of public and private development agencies, other clients of the construction industry, central and local government, and other regulatory authorities.		Descriptive and informative but of specialist nature not for general reading. The report assumes that readers will have a basic understanding of the nature of the problems of land contamination.
CIRIA Special Publication SP102 *Decommissioning, decontamination and demolition,* 1995	Section 2 paragraph 2.22 Section 4.3.4	Owners of contaminated sites. Non-specialist civil-engineering, architectural or construction advisers providing design, supervision and inspection services in collaboration with specialist advisers.	Deals with design and implementation of post-closure operations for contaminated sites.	

CIRIA Special Publication SP104 *Classification and selection of remedial methods*, 1995		Contracting organizations working on remediation projects. Regulatory bodies having responsibility for public and occupational health and safety in relation to a contaminated site.	Provides a classification of the techniques available for the treatment of contaminated sites.
CIRIA Special Publication SP105 *Excavation and disposal*, 1995			Deals with the excavation of contaminated material prior to disposal (on or off-site) or as a precursor to other forms of treatment.
CIRIA Special Publication SP106 *Containment and hydraulic measures*, 1996	Section 2 paragraph 2.22	Section 4.3.4	Owners of contaminated sites. Non-specialist civil engineering, architectural or construction advisers providing design, supervision and	Provides information and guidance on engineering-based remedial methods, specifically physical containment and hydraulic control measures.	Descriptive and informative but of specialist nature not for general reading. The report assumes that readers will have a basic

	understanding of the nature of the problems of land contamination.	
	inspection services in collaboration with specialist advisers. Contracting organizations working on remediation projects. Regulatory bodies having responsibility for public and occupational health and safety in relation to a contaminated site.	Deals with physical, chemical and biological methods of removing or rendering harmless the contaminants in solid materials after excavation from the ground.
CIRIA Special Publication SP107 *Ex-situ remedial methods for soils, sludges and sediments,* 1995		
CIRIA Special Publication SP109 *In-situ methods of remediation,* 1995		Describes the techniques available for removing, destroying or rendering harmless contaminants while they are in place in the ground.

Preparing for floods: interim guidance for improving the flood resistance of domestic and small business properties, DTLR, 2002	Section 2 paragraph 2.23	Section 4.4	Intended for use by property owners, developers, local planning authorities and others involved in construction of new buildings, and renovation of existing buildings, where their buildings are at risk of flooding.	The aim of the guide is to show how the flood resistance of properties may be improved.	Non-technical practical guidance.
BRE for Scottish Office Design guidance on flood damage for dwellings, TSO, 1996	Section 2 paragraph 2.23	Section 4.4	Of great use to anyone involved in the design, construction, renovation, repair and alteration of houses in flood risk area.	Covers issues affecting design, the effects of water on materials and guidance on construction and details.	A simple, eminently practical guide.

Assessment and management of risks to buildings, building materials and services from land contamination, Environment Agency 2001	Section 2 paragraph 2.24	Section 4.4	Aimed at all those involved in the management of land contamination including regulators, specialist consultants and contractors, construction clients, developers, main contractors, sub-contractors, consulting engineers and other construction professionals and producers of construction materials.	Considers the risks to buildings, building materials and services that may be associated with land contamination in terms of four principal hazards.	Descriptive and informative but requires some knowledge of risk analysis.
BRE Report BR 211 *Radon: guidance on protective measures for new dwellings* (third edition 1999)	Section 2 paragraph 2.40	Section 4.5.1	Essential reading for designers, contractors, regulators and developers.	Identifies those areas where either basic or full radon protection is needed by reference to a series of maps and provides practical guidance on measures to be taken.	Descriptive and informative non-technical guidance.

	Section 2 paragraph 2.40	Section 4.5.1	Statutory Instrument.	General appreciation of legal texts.
Ionising Radiations Regulations 1999 (SI 1999/3232)			Employers and self-employed persons responsible for a workplace.	
BRE Report BR 293 *Radon in the workplace*, 1995	Section 2 paragraph 2.40	Section 4.5.1	Provides guidance on radon protection for existing non-domestic buildings. It also includes information on the employer's responsibility under SI 1999/3232.	Descriptive and informative. Non-technical practical guidance.
		Aimed principally at employers and those who control buildings used for work purposes, or their representatives. Should also be of interest and assistance to surveyors and builders etc., concerned with specifying and carrying out the necessary remedial measures.		

Good Building Guide 25 *Buildings and radon*, 1996	Section 2 paragraph 2.40	Section 4.5.1	Essential reading for designers, contractors and house owners.	Provides interim guidance on radon measures in domestic buildings (including conservatories and extensions).	An easily readable guide with minimal technical jargon and plenty of simple illustrations.
HMIP Waste Management Paper No 27 *The control of landfill gas*, TSO, 2nd edition 1991	Section 2 paragraph 2.29	Section 4.6.4	Useful to anyone concerned with the development of land on or near landfill sites.	Discusses the main factors responsible for the formation of landfill gas in order to give information on the options available for its management.	Written in non-technical terms and is a useful way of gaining an understanding of the behaviour, monitoring and control of landfill gas.

Chartered Institution of Wastes Management (CIWM) entitled *Monitoring of landfill gas*, 2nd edition 1998	Section 2 paragraph 2.29	Section 4.6.4	Useful to anyone concerned with the development of land on or near landfill sites.	Provides detailed information on all aspects of gas generation and monitoring under a variety of conditions.	Descriptive and informative but more specialist than HMIP Waste Management Paper No 27.
Methane and other gases from disused coal mines: the planning response, TSO, 1996	Section 2 paragraph 2.29	Section 4.6.4	Provides advice suitable for use by planners, developers, land and property owners, insurers and others, in current and past coal mining areas.	Aimed at identifying a suitable planning response to reduce mine gas emission risks in respect of new development, without placing unnecessary constraints on land use.	Descriptive and informative, and non-technical.

CIRIA Report 130 *Methane: its occurrence and hazards in construction, 1993*	Section 2 paragraph 2.29	Section 4.6.4	Provides guidance for construction professionals who may have to take methane (and other gases often present with it) into account during the construction process.	Contains relevant information on methane and other gases including information which will enable construction professionals to recognize potential methane problems and to initiate the process of finding solutions for them.	Although there are some sections of general interest, much of the report requires advanced knowledge of chemistry and mathematics.
CIRIA Report 131 *The measurement of methane and other gases from the ground, 1993*	Section 2 paragraph 2.29	Section 4.6.4	Useful to specialist companies directly involved in the detection, measurement and monitoring of gases from the ground and for people responsible for commissioning their services.	Provides guidance on the detection, measurement and monitoring of gases in the ground.	Descriptive and informative but of specialist nature not for general reading.

CIRIA Report 150 *Methane investigation strategies*, 1995	Section 2 paragraph 2.29	Section 4.6.4	The report is of a specialist nature and will be of most use to those people responsible for commissioning methane investigations.	Contains guidance for good practice in the design and execution of site investigations for methane and associated gases in the ground.	Descriptive and informative but of specialist nature not for general reading.
CIRIA Report 151 *Interpreting measurements of gas in the ground*, 1995	Section 2 paragraph 2.30	Section 4.6.4	Useful to engineers allowing them to test the validity of gas measurements and their meaning.	Reviews the limitations of current gas measurement techniques and recommends ways to standardize and improve not only the techniques of measurement but also the ways to develop sound interpretations.	Contains highly technical information suitable for specialists.

CIRIA Report 152 *Risk assessment for methane and other gases from the ground*	Section 2 paragraph 2.30	Section 4.6.4	Mainly for the use of construction professionals, but also acts as a guidance document to assessors of risk of gas ingress in development situations. The non-technical parts will be of use to all those concerned with risk assessments of sites where the presence of ground gases is suspected.	Proposes a rational methodology for gas hazard evaluation and risk assessment appropriate for a wide range of construction situations and ground gases.	Descriptive and informative but of specialist nature not for general reading.
CIRIA Report 149 *Protecting development from methane*, 1995	Section 2 paragraph 2.36	Section 4.6.4	Of use for experts, practitioners, clients and owners of development, public utility engineers and those with a statutory or regulatory responsibility for development.	Examines the need for buildings to be protected from hazards arising from methane and other gases.	A practical guide to current accepted good practice. Descriptive and informative but of specialist nature not for general reading.

DETR/Arup Environmental PIT Research Report: *Passive venting of soil gases beneath buildings*, 1997 (in two volumes)	Section 2 paragraph 2.34	Section 4.6.4	For use by consultants, engineers and contractors who are engaged in the design of buildings on or near low gas potential sites. Also as a reference document to regulators.	Provides quantitative information on the relative performance of various ventilation media and guidance on the design of passive gas protective measures.	Highly technical and contains very little practical guidance.
BRE/ Environment Agency Report BR 414 *Protective measures for housing on gas-contaminated land*, 2001	Section 2 paragraph 2.34	Section 4.6.5	Essential reading for designers, contractors, regulators and developers when considering developing near landfill sites.	Contains practical guidance on construction methods to prevent the ingress of landfill gas into buildings.	Descriptive and informative. Non-technical practical guidance.

| Planning Policy Guidance Note PPG 25 *Development and flood risk*, DTLR, 2002 | Section 0 paragraph 0.8 | Section 5.1 | Regulators, professionals and developers dealing with possible development in areas at risk of flooding. | Planning guidance document which explains how flood risk should be considered at all stages of the planning and development process in order to reduce future damage to property and loss of life. | General appreciation of legal guidance documents. |
| CIRIA publication C506 *Low cost options for prevention of flooding from sewers*, 1998 | Section 3 paragraph 3.6 | Section 5.2 | Aimed at the needs of drainage engineers and planners working for sewerage undertakers, local authorities, contractors and developers. | Summarizes the results of a CIRIA project in which low cost options for preventing flooding from sewers were identified and information collected on their suitability and effectiveness. | Descriptive and informative but of specialist application and not for general reading. |

CIRIA/ Environment Agency Flood products. Using flood protection products – a guide for home owners, 2003	Section 0 paragraph 0.8	Section 5.2	Aimed specifically at home owners (and tenants, presumably).	Designed to help its target audience to assess the risk that flooding entails and to understand the different routes through which floodwater may enter.	A handy non-technical guide of universal use in areas prone to flooding.
BRE Digest 276 *Hardcore*, 1992	Section 4 paragraph 4.7	Section 6.3.1	Like most other BRE Digests, this one is of universal application to all concerned with the built environment.	Provides information on hardcore, considers the suitability of some of the materials in common use and discusses some of the problems that can arise.	Descriptive, informative and non-technical.

BS 8500-1: 2002 *Concrete. Complementary British Standard to BS EN 206-1. Method of specifying and guidance for the specifier*	Section 4 paragraph 4.7	Section 6.3.1	Essential reading for all people concerned with the specification and ordering of concrete for use on site.	Offers guidance to the specifier on five approaches to the specification of concrete.	Specialist document dealing with concrete specification.
BS 1282: 1999 *Wood preservatives Guidance on choice, use and application*	Section 4 paragraph 4.11	Section 6.3.1	Useful to anyone concerned with the treatment of wood in the built environment such as architects, building surveyors, specialist timber treatment companies and contractors.	Provides an overview of wood preservation and the factors for consideration in the selection, and application of appropriate wood preservatives and in the use of preservative-treated timber.	Descriptive, informative and non-technical.

CP 102: 1973 *Protection of buildings against water from the ground* (Clause 11)	Section 4 paragraph 4.12	Section 6.3.1	Of particular use to designers, surveyors and contractors engaging in renovation and repair work. Also of use to students.	Deals with the methods of preventing the entry of ground water and surface water into a building from the surrounding areas.	Well illustrated and easy to read.
BS 8102: 1990 *Code of practice for protection of structures against water from the ground*	Section 4 paragraph 4.12	Sections 6.3.1 and 7.3	Of use to designers, general and specialist contractors, and regulatory authorities.	Provides guidance on methods of dealing with and preventing the entry of water from surrounding ground into a building below ground level.	Well illustrated and easy to read.

BRE Digest 429 *Timbers and their natural durability and resistance to preservative treatment*	Section 4 paragraph 4.15	Section 6.3.2	Useful to anyone concerned with the treatment of wood in the built environment such as architects, building surveyors, specialist timber treatment companies and contractors.	Contains an explanation of the classification of durability and treatability for timber.	Descriptive, informative and non-technical.
BS 7331: 1990 *Specification for direct surfaced wood chipboard based on thermosetting resins*	Section 4 paragraph 4.15	Section 6.3.2	Of particular use to people concerned with the specification and detailed design of timber floors.	Specifies the requirements for six types of direct surfaced wood chipboard for interior applications.	Descriptive and informative.

BS EN 312 Part 5 *Particleboards. Specifications. Requirements for load-bearing boards for use in humid conditions*: 1997	Section 4 paragraph 4.15	Section 6.3.2	Of particular use to people concerned with the specification and detailed design of timber floors.	Specifies the requirements for mechanical and swelling properties of the boards and contains requirements for moisture resistance.	Descriptive and informative.
BS 5250: 2002 *Code of practice for the control of condensation in buildings.* Appendix D.	Section 4 paragraph 4.21	Section 6.3.4	Of use to building designers, contractors, owners, managers and occupiers.	Describes the causes and effects of surface and interstitial condensation in buildings and gives recommendations for their control.	Descriptive and informative.

BS EN ISO 13788: 2001 *Hygrothermal performance of building components and building elements. Internal surface temperature to avoid critical surface humidity and interstitial condensation. Calculation methods*	Section 4 paragraph 4.21	Section 6.3.4	Mainly aimed at building design specialists with the required level of knowledge of physics and higher mathematics.	Gives calculation methods for: • the internal surface temperature of a building component or building element below which mould growth is likely • the assessment of the risk of interstitial condensation due to water vapour diffusion.	See target readership. Not for general reading.
BRE Report BR 262 *Thermal insulation: avoiding risks,* 2002			As with most of the BRE's publications this report should form part of the reference library of all those concerned with the design, construction and use of buildings.	Discusses the more important technical risks associated with meeting the requirements of building regulations for thermal insulation.	Descriptive, informative and non-technical. Well illustrated.

Limiting thermal bridging and air leakage: robust construction details for dwellings and similar buildings, published by The Stationery Office, 2002. (in 8 volumes)	Section 4 paragraph 4.22	Section 6.3.5	Essential reading and of great practical use to both designers and contractors. Essential for building control bodies.	Details prepared to assist the construction industry in achieving the performance standards published in the Building Regulations Approved Documents L1A, L1B, L2A and L2B (2006 editions).	Descriptive, informative and non-technical. Well illustrated.
BRE Information Paper IP17/01 *Assessing the effects of thermal bridging at junctions and around openings*	Section 4 paragraph 4.22	Section 6.3.5	Will only be of use to specialists dealing with energy use in buildings and since publication of the 2006 edition of Approved Document L (4 volumes) is now out of date.	Gives guidance on assessing the effects of thermal bridging at junctions and around openings in the external elements of buildings and how to assess their effect on the overall heat loss.	Assumes readers understand the principles, and are familiar with the calculation, of heat loss through the external elements of buildings.

BS 8215: 1991 *Code of practice for design and installation of damp-proof courses in masonry construction.* Clauses 4 and 5	Section 5 paragraph 5.6	Section 7.3	Essential reading for all people concerned with the specification and detailed design of buildings.	Contains recommendations for the selection, design and installation of damp-proof courses (DPCs) in both solid and cavity masonry constructions.	Descriptive, informative and non-technical. Well illustrated.
BS 8104: 1992 *Code of practice for assessing exposure of walls to wind-driven rain*	Section 5 paragraph 5.8	Section 7.4.1	Useful for designers and specifiers when anticipating the possible use of solid wall construction in masonry.	Gives recommendations for two methods for assessing exposure of walls in buildings to wind-driven rain (i.e. the local spell index method and the local annual index method).	Relatively easy to use and contains in appendix A, a number of worked examples.

BS EN 998: *Specification of mortar for masonry Part 2: 2002: Masonry mortar*	Section 5 paragraph 5.9	Section 7.4.1	Specifies requirements for masonry mortars (bedding, jointing and pointing) for use in masonry.	Essential reading for all people concerned with the specification and detailed design of masonry in buildings.	Descriptive and informative.
BS 5262: 1991 *Code of practice for external renderings*	Section 5 paragraph 5.9	Section 7.4.1	Gives recommendations for cement-based external renderings on all common types of background.	Invaluable to anyone concerned with the specification and detailing of building work involving rendering.	Easily readable, descriptive and informative.
BS 5628 *Code of practice for use of masonry Part 3: 2001 Materials and components, design and workmanship*	Section 4 paragraph 4.14 and Section 5 paragraph 5.8	Section 6.3.2, 7.4.1 and 7.4.2	Part 3 of the BS gives general recommendations for the design, construction and workmanship of masonry, including materials and components and the main aspects of design.	Essential reading for all people concerned with the specification and detailed design of buildings.	Descriptive, informative and non-technical. Well illustrated.

BS5617: 1985 *Specification for urea-formaldehyde (UF) foam systems suitable for thermal insulation of cavity walls with masonry or concrete inner and outer leaves*	Section 5 paragraph 5.15	Section 7.4.3	Due to the problems that have been encountered with this product in the past it is seldom used today as a cavity insulant.	Specifies the property requirements, the properties of the components and the production parameters, of urea-formaldehyde foam systems suitable for injection into external masonry or concrete cavity walls to provide improved thermal insulation.	See note under target readership.
BS5618: 1985 *Code of practice for thermal insulation of cavity walls (with masonry or concrete inner and outer leaves) by filling with urea-formaldehyde (UF) foam systems*	Section 5 paragraph 5.15	Section 7.4.3	Due to the problems that have been encountered with this product in the past it is seldom used today as a cavity insulant.	Describes recommendations for the installation of UF foam systems which are dispensed on site, to fill the cavities of suitably situated and constructed external walls of maximum height 12 m, thereby providing additional thermal insulation to such walls.	See note under target readership.

| BS 8208: Part 1: 1985 *Guide to suitability of external cavity walls for filling with thermal insulation* | Section 5 paragraph 5.15 | Section 7.4.3 | Intended for people who are appropriately qualified and experienced and are properly trained to carry out assessments of cavity walls for filling with insulants. | Gives guidance on factors to be considered when assessing the suitability of existing external cavity walls with masonry and/or concrete leaves for filling with thermal insulants. | Specialist document not for general reading. |
| BS 8000: Part 6: 1990 *Workmanship on building sites. Code of practice for slating and tiling of roofs and claddings* | Section 5 paragraph 5.25 | Section 7.4.4 | Essential reading for all people concerned with the specification, detailed design and construction of slate and tile roofs and claddings. | This part of BS 8000 gives recommendations on basic workmanship and covers those tasks which are frequently carried out in relation to slating and tiling of roofs and claddings of buildings. | Descriptive and informative but lacking illustrations. |

British Standard Code of Practice 143 *Code of practice sheet roof and wall coverings (various parts)*	Section 5 paragraph 5.28	Section 7.4.5	Essential for anyone concerned with the design, construction, repair and maintenance of sheet roof and wall coverings in traditional materials.	The code consists of a series of parts dealing with different sheet roofing and walling materials.	Easy to read and contain numerous drawings and details to illustrate the design and construction principles.
BS 6915: 2001 *Specification for design and construction of fully supported lead sheet roof and wall coverings*	Section 5 paragraph 5.28	Section 7.4.5	Essential for anyone concerned with the design, construction, specification, repair and maintenance of lead sheet roof and wall coverings.	Gives recommendations for the design and construction of fully supported coverings of rolled lead sheet, conforming to BS EN 12588, applied to external wall and roof surfaces.	Easy to read document containing a large number of practical details and much useful advice.

	Section 5 paragraph 5.28	Section 7.4.5			
BS 8219: 2001 *Profiled fibre cement. Code of practice*	Section 5 paragraph 5.28	Section 7.4.5	Essential for anyone concerned with the design, construction, repair and maintenance of fibre cement sheet roof and wall coverings.	Gives recommendations for design specific to the use of profiled fibre cement sheets for roof and wall cladding on buildings, and recommendations for basic workmanship and tasks carried out in relation to the installation of profiled fibre cement sheets.	Easy to read document containing a large number of practical details and much useful advice.
BS 8200: 1985 *Code of practice for design of non-loadbearing external vertical enclosures of buildings*	Section 5 paragraph 5.28	Section 7.4.5	Intended to be used primarily by designers to assist in the design process. Will also assist designer in his role of supervisor,	Provides a systematic framework within which an enclosure can be designed and constructed.	Specialist document not for general reading.

			assessor and advisor to his client and will be of use manufacturers, erector and main contractor.		Specialist document but containing a large number of practical details and much useful advice.
BS 8297: 2000 *Code of practice for design and installation of non-loadbearing precast concrete cladding*	Section 5 paragraph 5.28	Section 7.4.5	Essential for designers, contractors, manufacturers and regulators concerned with the use of non-load-bearing precast concrete panel construction.	Gives recommendations and guidance for the design, manufacture, transport and installation of non-load-bearing precast concrete cladding in the form of: • units supported by and fixed to a structural frame or wall, and • units used as permanent formwork in part or in whole.	

	Section 5 paragraph 5.28	Section 7.4.5			
BS 8298: 1994 *Code of practice for design and installation of natural stone cladding and lining*	Section 5 paragraph 5.28	Section 7.4.5	Essential for all concerned with the design, installation and maintenance of mechanically fixed facing units of natural stone.	Contains recommendations for the design, installation and maintenance of mechanically fixed facing units of natural stone as a cladding and lining where these are: • held to a structural background by metal fixings, or • attached to precast concrete units (i.e. stone faced concrete cladding units).	Specialist document but practical and informative.
MCRMA Technical Paper 6 *Profiled metal roofing design guide,* revised edition 2004	Section 5 paragraph 5.28	Section 7.4.5	Of use to all concerned with the design, construction and regulation of profiled metal roofing.	Applies to profiled steel or aluminium sheets used in various roof constructions such as single skin and double skin	This excellent paper provides non-technical and highly readable information.

	Section			
			systems, secret fix, site assembled or factory made composite panels and under purlin linings.	This paper provides non-technical and highly readable information, however, it is now well out of date, especially regarding thermal performance.
	Section 7.4.5	Of use to designers and contractors in giving an indication of the general principles for the design and installation of these products, but should not be used for thermal design as it is well out of date.		
MCRMA Technical Paper 9 *Composite roof and wall cladding panel design guide,* 1995	Section 5 paragraph 5.28		Covers the materials, methods of manufacture and performance of metal composite roof and wall cladding panels.	

	Section 5 paragraph 5.18	Section 7.5		
BRE Building Elements. Walls, windows and doors (Performance, diagnosis, maintenance, repair and the avoidance of defects) 1998		Primarily aimed at building surveyors and other professionals performing similar functions (such as architects and builders) who maintain, repair, extend and renew the national building stock. It will also be of use to students.	Covers all kinds of external walls, both loadbearing and non-load-bearing; windows and doors. Describes both good and bad features of walls, windows and doors and the joints between them. Concentrates on those aspects of construction which, in the experience of BRE, lead to the greatest number of problems or greatest potential expense, if carried out unsatisfactorily.	An excellent, well illustrated and informative book.

BRE Report BR 292 *Cracking in buildings: 1996*	Section 5 paragraph 5.18	Section 7.5	Aimed at three main interest groups: • Construction professionals (architects, surveyors and contractors) • Litigators (building failure investigators, loss adjusters and expert witnesses) • building owners and maintenance staff.	Describes the science behind the different causes of cracking thereby enabling surveyors and others who investigate the phenomenon to understand the causes and hence be able to offer diagnoses and remedies.	Highly readable and well illustrated. Set out so as to be readily accessible to all members of the groups opposite. Does require some knowledge of basic chemistry and physics.
BRE Good Building Guide 47 *Level external thresholds: reducing moisture penetration and thermal bridging, 2001*	Section 5 paragraph 5.33	Section 7.7	Mostly intended for practitioners to encourage and improve mutual awareness of the roles of different trades and professions.	Describes some of the technical risks associated with the design of level thresholds and some detailing solutions.	Well illustrated and written in clear non-technical terms.

Accessible thresholds in new buildings: guidance for house builders and designers. TSO, 1999	Section 5 paragraph 5.33	Section 7.7	Essential for all who are concerned with the design, construction and regulation of dwellings.	The aim of this guide is to: • suggest design solutions that will make dwelling thresholds more accessible to disabled people, whilst minimising the risk of water entering the building, and • help designers achieve solutions that do not conflict with other aspects of Building Regulations.	Well illustrated and written in clear non-technical terms.

| MCRMA Technical Paper 14 *Guidance for the design of metal roofing and cladding to comply with approved document L2: 2001, 2002 revision* | Section 6 paragraph 6.14 | Section 8.5 | This edition is no longer applicable. | This publication was originally intended to assist designers, manufacturers and installers of metal walls and roofs to comply with the requirements of Part L of the Building Regulations which came into force in April 2002. These have now been superceded thereby rendering much of this publication obsolete, although the sections on infra-red surveys and air leakage testing are of some general interest. | No longer applicable. |

Index

(Page numbers for figures have suffix **f**, those for tables have suffix **t**)